W0107179

Springer Japan KK

E.R. Nakamura (Editor-in-Chief)
K. Kudo, O. Yamakawa, Y. Tamagawa (Editors)

Complexity and Diversity

With 192 Figures

Springer

Editor-in-Chief:
Eiichi Ryoku Nakamura
Department of Bioscience, Fukui Prefectural University, 4-1-1 Kenjojima, Matsuoka-cho, Yoshida-gun, Fukui, 910-11 Japan

Editors:
Kiyoshi Kudo
Department of Applied Physics, Fukui University, 3-9-1 Bunkyo, Fukui, 910 Japan

Osamu Yamakawa
Center of Information Science, Fukui Prefectural University, 4-1-1 Kenjojima, Matsuoka-cho, Yoshida-gun, Fukui, 910-11 Japan

Yoichi Tamagawa
Department of Applied Physics, Fukui University, 3-9-1 Bunkyo, Fukui, 910 Japan

We would like to acknowledge that the symbol of "Dissertation de Arte Combinatoria" on the cover was also used in the Japanese edition of *Gottfried Wilhelm Leibniz, OPERA OMNIA*, vol. 1, published by Kousakusha, Japan.

ISBN 978-4-431-66864-0   ISBN 978-4-431-66862-6 (eBook)
DOI 10.1007/978-4-431-66862-6

Printed on acid-free paper

© Springer Japan 1997
Originally published by Springer-Verlag Tokyo in 1997
Softcover reprint of the hardcover 1st edition 1997
This work is subject to copyright. All rights are reserved, whether the whole or part of the material is concerned, specifically the rights of translation, reprinting, reuse of illustrations, recitation, broadcasting, reproduction on microfilms or in other ways, and storage in data banks.
The use of registered names, trademarks, etc. in this publication does not imply, even in the absence of a specific statement, that such names are exempt from the relevant protective laws and regulations and therefore free for general use.

Typesetting: Camera-ready by authors

Preface

Interdisciplinary studies of complex systems extend across a wide range of scientific disciplines: applied mathematics, physics, chemistry, biology, psychology, ecology, sociology and economics. In nonlinear complex open systems, diversity is a prominent feature in the process of self-organization based on nonlinear interactions of abundant elements.

It is an open question whether a great diversity of complex systems can be explained by a scientific paradigm. Interdisciplinary studies should play the essential role in answering this question.

The aim of this book is to study the origins of the diversity of complex systems. A strategy to clarify the relation between complexity and diversity will contribute to investigations of central dynamics in complex systems. This volume is divided into four parts according to the latest research on complex systems: mathematical systems, physical systems, living systems, and social systems.

The contributions to this volume are based on presentations at the Complexity and Diversity workshop which was held at the Center for Academic Exchange of Fukui Prefectural University in Fukui, Japan, August 28–30, 1996. In addition to the material included in this volume, talks were also given by the following prominent scientists: Prof. T. Tada, Prof. Y. Shiozawa, and Prof. K. Nishiyama. The article titled "Synergetics and Complexity: Emerging New Science and Civilization at the Turn of the Century" was contributed by Prof. K. Mainzer especially for this book.

I want to express my gratitude to Prof. K. Sakamoto, President of Fukui Prefectural University, and to Prof. A. Ueda. This workshop was held with the financial support of the Fukui Prefecture Universities Research Foundation for the Promotion of Sciences and the Fukui Convention Bureau.

Fukui, Japan
January 1997

E. R. Nakamura

Contents

Part III   Physical Systems

Part IV   Living Systems

Part V   Social Systems

Part I Overview

Complexity and Diversity – An Introduction

E.R. Nakamura[1] and K. Kudo[2]

1 Department of Bioscience, Fukui Prefectural University, Matsuoka, Fukui 910-11, Japan

2 Department of Applied Physics, Fukui University, Fukui 910, Japan

Abstract

Measures of complexity hitherto proposed are classified into two, i.e. first order complexity C_1 and second order complexity C_2. It is difficult to comprehend complexity of self-organization by means of these measures. We have proposed a way to analyze the mechanism of self-organization on the basis of the convolution model to make the relation between the states of total system and subsystems. The state of total system is convoluted by the states of subsystems in some phase space, while the state of subsystem is given by convolution of the states of total system as the feedback effect in the process of time development. It has been pointed out that complexity of self-organization results from antagonistic relationality between total system and subsystems to set these systems in the respective appropriate states.

Keywords: Complexity, diversity, self-organization, relationality

1. What is complexity?

In the process of construction of various dynamical theories the view of reductionism to reduce natural phenomena to some simple elements and interactions played an essential role for our recognition of the objective world [1]. Most of the scientists believed that every behaviour of the system, which was composed of these elements, was symmetric in the space-time transformation and was perfectly deterministic. Their belief for some ideal object reminds us of the Pythagorean doctorine to hallow the perfect circle.

Every simple system is isolated and closed in a naive sense. This is the reason why the system preserves some time independent quantities provided by the conservation law. Characteristic fearures of a simple system can be deduced from linear interaction between elements of the system. If the equation of motion of this system yields some solutions, they must be superposed according to the principle of symmetry. Model building of such a simple system has enabled the old scientists to explore dynamical world until the 19th century.

However, it is not difficult to comprehend that we have obtained the view of simple systems from ideal models neglecting various effects from surroundings. The dissipative system in the real world is not closed, and hence there are some flows of energy, information, etc., between the system and its environment. We can imagine usual frictional phenomena of the dissipative system which is not simple but rather complex [2]. Linear interaction in the dynamical model is also our ideal assumpion to desire some symmetric world which is demanded from the mathematical beauty or simplicity. It is well known, however, that in the ordinary development of atmospheric phenomena and liquid flow one often finds the chaotic singularity of these phenomena like vortex. This singularity, which breaks regular development of simple systems, is caused by nonlinear interaction among elements. When we deal with the nonlinear open system, we must abandon the view of simple system and think out an appropriate strategy to explore the structure of complex systems, e.g. nerve systems, brain systems, ecological systems, and social systems.

Neither odered state nor disordered state, which can be represented by usual dynamical models, show any features of complex state. Characteristics of complete disordered random state can be reproduced by some models with the use of maximum entropy, while ordered state should be given by order parameters. Maximum entropy and order parameter are good measures of simple states. As concerns complex state, how can we define the measure of complexity and estimate its quantity?

Algorithmic complexity has been estimated as the computational task. Algorithmic complexity like Kolmogorov complexity is defined by $C_1 \sim \log N$ in terms of N digits of the sequence set. This definition of complexity represents the increasing number of sequential events. We will call C_1 as *first order complexity*. In random state C_1 becomes large as entropy increases.

It should be noticed that change of *first order complexity* ΔC_1 in the large N region is nearly equal to zero. If we use Shannon entropy instead of $\log N$ and sum up $\Delta C_1(N)$ (change of Shannon entropy) with respect to N according to Grassberger's stored information (Grassberger's effective complexity [3]), we obtain the expression of complexity $C_2 = \sum_N N(\Delta C_1(N-1) - \Delta C_1(N))$ concerning stored information for some sequence set. Complexity C_2 may be called *second order complexity*.

We can characterize systems in random state with the use of maximum entropy, while we apply order parameter to systems in ordered state. It has been stressed that effective complexity should be high only in the intermediate region between random and order. However, the value of *second order complexity* like effective complexity is large only in the case of that Shannon entropy changes regularly in oder to store necessary informations. If a system shows some irregular changes to adapt itself to various surroundings, stored information of the system does not always become large. Thus, effective complexity is not appropriate to the system which yields the irregular change like self-organization.

High effective complexity does not indicate high self-organization, and hence does not produce high diversity. Therefore, we must explore complexity with respect to self-organization, since self-organization leads to diversity of the system according to the change of surroundings.

2. Dynamics of self-organization

Physicists hitherto have investigated self-organization in comparison with self-ordering of various physical systems in the framework of statistical mechanics. As a typical example we can refer to Haken's slaving principle [4] through order-disorder phase transition.

Cluster formation, which yields long range correlation between elements in odered phase, plays an essential role in the model of order-disorder phase transition. If we utilize recurrent cluster formation according to the renormalization group method in statistical mechanics, we arrive at the result that effective coupling of interaction between elements approaches to a critical value in ordered phase.

Cluster formation and the slaving effect in order-disorder phase transition result from non-linear interaction among elements. We can deduce this singular behaviour from a convolution model [5] as a result of the nest-structure with respect to stratum of clusters. According to this model, the state density of a system σ is constructed by convolution of state densities of its subsystems ρ_i through multiple integral in some phase space as $\sigma(N) = \rho_1 * \rho_2 * \cdots * \rho_N$. In the limit of scale invariance it turns out that we obtain fractal structure of macroscopic variable of state like correlation length ξ in terms of step number N, i.e. $\xi = N^{1/D}$, where D is fractal dimension.

We can get macroscopic variable of state of a system through convolution of various states of subsystems. In the ordinary Hamiltonian formalism, where variables of state are conserved, time independent local interaction in subsystems produces variable of state of the total system which leads to order-disorder phase transition at the critical point. However, in the process of self-organization the system must change local interaction in subsystems to adapt itself to surroundings. This indicates that the condition of mimimum free energy is not always valid because of fluctuation of energy in the adaptation process. Time dependent state density of subsystems yields various states of the total systems accoding to convolution of subsystems. If the condition of mimimum free energy is not valid, it may be difficult to decide definitely the state of total system from local interaction in subsystems.

The convolution formula provides only the relation of the states between total system and subsystems. In contrast with self-ordering phenomena, where variable of the state of total system is given by local interaction in subsystems, self-organization ought to controll interation in subsystems on the basis of the feedback effect from the time dependent states of total system. We can represent the feedback mechnism from the states of total system to the state of subsystem in the form of convolution by means of multiple integral through time development like $\rho(t) = \sigma_{t_1} * \sigma_{t_2} * \cdots * \sigma_{t_N}$. This convolution formula indicates that time dependent local interaction in subsystem is affected by the historical states of total system σ_{t_i} during time interval t_i.

We cannot account complexity of self-organization on the basis of simple interaction between elements. It is appropriate to consider that this complexity results from complex characteristics of the system [6]. In the convolution model the self-organization mechanism can be represented by the convolution relation between total system and subsystems through the feedback effect according to time development. Divergent states of the system reflect its historical development in terms of time t_N which should be complex variable. Under the condition of scale invariance, if we set $t_N = \log N$ in the form of *first order complexity*, we get fractal structure of distance of two nearby trajectories like $d(t) = e^{\lambda t}$, where λ corresponds to the Lyapunov exponent of the chaotic system.

How can we measure complexity like self-organization? In this article we have pointed out relationality between total system and subsystems based on the convolution model. According to the slaving principle, one may imagine antagonistic relation between states of two systems in the process of time development. As concerns antagonistic relation, we can show various examples of complex self-organization like recognition and learning. In the following section we will propose our attempt to measure complexity of recognition by means of a model simulation.

3. Pattern recognition with degree of satisfaction

By ordinary we are satisfied with the information which is new and rather easy to understand. If the information is already known or too difficult to understand, we have little interest in it. Therefore, there should exist the state of maximum satisfaction between two antagonistic informations, i.e. trivial information and difficult information. We apply this idea to pattern recognition introducing parameter α for decrease in understanding. In pattern recognition we notice another two antagonistitic pattern matchings, i.e. global pattern matching which means adaptation to surroundings and local pattern matching which search appropriate local interaction. We, therefore, investigate the balance between global and local patterns by using a mixing parameter λ.

In the pattern recognition process we deal with two sets of antagonistic relations by using parameter α as pattern scrambling with respect to understanding and global-local mixing parameter λ. Let us show the model calculation in the following;

1. Select four global references $|g_i\rangle$ $(i = 1 \sim 4)$ in bit image.

2. Divide $|g_i\rangle$ into nine blocks $|l_i\rangle$ $(i = 1 \sim 9)$. $|l_i\rangle$ made of nine pixels are used as local references.

3. Make original tiling pattern from global references $|g_i\rangle$, next contaminate it randomly and create scrambled pattern $|S_i\rangle$ with fractional noise parameter α. $\alpha = 0$ means no change, and $\alpha = 1$ gives whole reverse pattern.

4. Construct global tiling pattern $|G_i\rangle$ by patching global references $|g_i\rangle$ and local tiling pattern $|L_i\rangle$ by patching local references $|l_i\rangle$ so as to give the lowest value to the following cost function F;

$$F = (1 - \lambda) \sum_i \langle L_i | G_i \rangle + \lambda \sum_i \langle L_i | S_i \rangle,$$

where $\langle L_i | G_i \rangle$ $(\langle L_i | S_i \rangle)$ means global (local) Hamming distance of local tiling pattern from global tiling (scrambled) pattern.

Making use of control parameter λ, we compromise global pattern matching with local pattern matching. Another parameter α represents the distance of scrambled pattern from original global tiling pattern. Thus, α presents another compromise in terms of degree of satisfaction. Let us define a new measure of complexity C, which gives the measure of compromise with respet to antagonistic relations, in terms of the distribution of F as follows;

$$C = \langle F^2 \rangle - \langle F \rangle^2.$$

The measure C means degree of satisfaction for pattern recognition. We can make original pattern from four global references in the model simulation. Fig. 1. shows that the distribution C has the highest value near $\alpha = 0.35$ and $\lambda = 0.35$. In our pattern recognition model we conclude that we are inclined to watching global patterns rather than local reference patterns, and we satisfy when the noise is 35%.

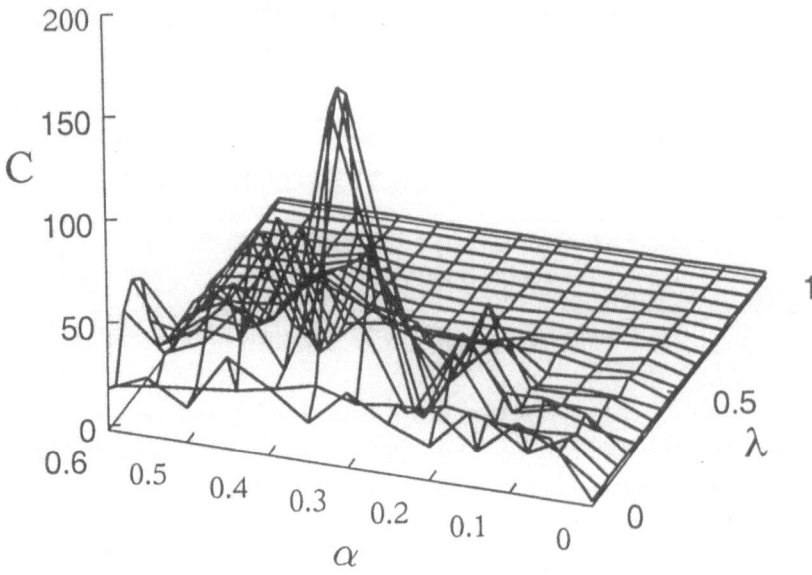

Fig.1 The degree of satisfaction C versus noise parameter α and global-local mixing parameter λ.

4. Complexity and relationality

With respect to the phenomena of self-organization we acknowledge two antagonistic tendencies, i.e. the tendency toward adaptation to new surroundings and the tendency to maintain the system in the same state. In the previous section we have shown an example of these characteristics in the process of recognition by means of antagonistic matchings for global and local patterns. In the process of pattern matching we can comprehend that global matching for objective target pattern means adaptation to surroundings, while local matching is to search reliable local interaction. We thus obtain divergent patterns according to various patterns given by induced noises.

It may be possible to point out that complexity in the process of recognition indicates the degree of our interest. Both of complete agreement and disagreement in matchings, where antagomistic relation breaks, decrease our interest for target patterns, and hence complexity of pattern matching becomes low. It turns out from high antagonistic relationality of matchings between global and local patterns that the system to pursue pattern matching yields high complexity.

High antagonistic relationality is an important concept of ecological and social systems. Relationality between total system and subsystems shall play an important role in self-organzation of these systems, too. It may be challenging task to explore antagonistic relationality in various systems in more detail.

References

[1] S. Weinberg, *Dreams of a Final Theory*, Pantheon Books, New York, 1992.

[2] G. Nicolis, I. Prigogine, *Exploring Complexity: An Introduction*, Freeman & Co., München, 1989.

[3] P. Grassberger, How to Measure Self-Generated Complexity, *Physica*, **140a**, 1986, pp. 319-325.

[4] H. Haken. *Synergetics-An Introduction*, 3rd Edn., Springer-Verlag, Berlin, 1983.

[5] E.R. Nakamura, K. Kudo, A Convolution Formula for a Phase Transition of Hadronic Matter and the Multiplicity Distributions, *Physics Letters*, **B191**, 1988, pp. 381-385. Convolution Theory of a Phase Transition between Hadronic and Quark Matter and the Characteristic Multiplicity Distriutions, *Physical Review*, **D40**, 1990, pp. 281-284.

[6] M. Gell-Mann, What is Complexity?, *Complexity*, **1**, 1995, pp. 16-19.

SYNERGETICS AND COMPLEXITY:
EMERGING NEW SCIENCE AND CIVILIZATION
AT THE TURN OF THE CENTURY

Klaus Mainzer

Abstract: The theory of nonlinear complex systems has become a successful problem solving approach in the natural sciences of physics, chemistry and biology. It is now recognized that many of our social, economical, ecological, and physical problems are also of a global, complex, and nonlinear nature. And one of the most exciting contemporary topics is the idea that even the human brain and mind is governed largely by the nonlinear dynamics of complex systems. Synergetics aims at mathematical and computational models with nonlinear dynamics and well-defined order parameters. These methods are sometimes said to foreshadow the new sciences of complexity characterizing the scientific development of the 21st century [1].

1.   From Linear to Nonlinear Thinking

What is the reason behind the successful interdisciplinary applications of nonlinear complex systems? This approach cannot be reduced to special natural laws of physics, although its mathematical principles were discovered and at first successfully applied in physics. Thus it is no kind of traditional 'physicalism' to explain the dynamics of laser, ecological populations, or our brain by similar structural laws. It is an *interdisciplinary methodology* to explain the *emergence of certain macroscopic phenomena* via the *nonlinear interactions* of *microscopic elements* in *complex systems*. Macroscopic phenomena may be forms of light waves, fluids, clouds, chemical waves, plants, animals, populations, markets, and cerebral cell assemblies which are characterized by order parameters. They are not reduced to the microscopic level of atoms, molecules, cells, organisms, etc., of complex systems. Actually, they represent properties of real macroscopic phenomena, such as field potentials, social or economical power, feelings or even thoughts. Who will deny that feelings and thoughts can change the world?

In history the concepts of the social sciences and humanities have often been influenced by physical theories. In the age of mechanization Thomas Hobbes described the state as a machine ('Leviathan') with its citizens as cog wheels. For Lamettrie the human soul was reduced to the gear drive of an automaton. Adam Smith explained the mechanism of the market by an 'invisible' force like Newton's gravitation. In classical mechanics causality is deterministic in the sense of the Newtonian or Hamiltonian equations of motion. A conservative system is characterized by its reversibility (i.e., symmetry or invariance) in time and the conservation of energy. Celestial mechanics and the pendulum without friction are prominent examples. Dissipative systems are irreversible, like Newton's force with a friction term, for instance.

But, in principle, nature was regarded as a huge *conservative* and *deterministic* system the causal events of which can be forecast and traced back for each point of time in the future and past if the initial state is well known ('Laplace's demon'). It was Henri Poincaré who recognized that celestial mechanics is no completely calculable clockwork even with the restrictions of conservation and determinism. The causal interactions of all planets, stars, and celestial bodies are *nonlinear* in the sense that their mutual effects can lead to chaotic trajectories (e.g., the 3-body problem). Nearly sixty years after Poincaré's discovery, A.N. Kolmogorov (1954), V.I. Arnold (1963), and J.K. Moser proved the so-called KAM theorem: Trajectories in the phase space of classical mechanics are neither completely regular nor completely irregular, but they depend very sensitively on the chosen initial states. Tiny fluctuations can cause chaotic developments (the 'butterfly effect').

In this century quantum mechanics has become the fundamental theory of physics. The superposition or linearity principle of quantum mechanics delivers correlated ('entangled') states of combined systems which are highly confirmed by the EPR experiments (A. Aspect 1981). But a quantum field equation with a two-particle potential, for instance, contains a nonlinear term corresponding to pair creation of elementary particles. In general the reactions of elementary particles in quantum field theory are essentially nonlinear phenomena. The interactions of an elementary particle cause its quantum states to have only a finite duration and thereby to violate the reversibility of time. Thus even the quantum world itself is neither conservative nor linear in general.

2.   Synergetics and Complex Natural Systems

Since the presocratics it has been a fundamental problem of natural philosophy to discover how order arises from complex, irregular, and chaotic states of matter. Heraclitus believed in an ordering force of energy (*logos*) harmonizing irregular interactions and creating order states of matter. Modern thermodynamics describes the emergence of order by the mathematical concepts of statistical mechanics. We distinguish two kinds of phase transition (self-organization) for order states: *Conservative self-organization* means the phase transition of reversible structures in or near to the thermal equilibrium. Typical examples are the growth of snow crystals or the emergence of magnetisation in a ferromagnet by annealing the system to a critical value of temperature. Conservative self-organization mainly creates order structures with low energy at low temperatures. Nanostructures such as Buckminsterfullerene forming large balls of carbon molecules are examples of chemical self-assemblies near to thermal equilibrium. *Dissipative self-organization* is the phase transition of irreversible structures far from thermal equilibrium. Macroscopic patterns arise from the complex nonlinear cooperation of microscopic elements when the energetic interaction of the dissipative ('open') system with its environment reaches some critical value. The stability of the emergent structures is guaranteed by some balance of *nonlinearity* and *dissipation*. Too much nonlinear interaction or dissipation would destroy the structure.

As the conditions of dissipative phase transition are very general, there is a broad variety of interdisciplinary applications. A typical physical example is the laser. In chemistry, the concentric rings or moving spirals in the Belousov-Zhabotinski (BZ) reaction arise when specific chemicals are poured together with a critical value. The competition of the separated ring waves show the nonlinearity of these phenomena very clearly, because in the case of a superposition principle the ring waves would penetrate each other like optical waves.

The phase transitions of nonlinear dissipative complex systems are explained by synergetics. In a more qualitative way we may say that old structures become unstable and break down by changing *control parameters*. On the microscopic level the stable modes of the old states are dominated by unstable modes (Haken's *'slaving principle'*). They determine *order parameters* which describe the macroscopic structure and patterns of systems. There are different final patterns of phase transitions corresponding to different attractors. Different attractors may be pictured by a stream, the velocity of which is accelerated step by step. At the first level a homogeneous state of equilibrium is shown ('fixed point'). At a higher level of velocity the bifurcation of two or more vortices can be observed corresponding to periodic and quasi-periodic attractors. Finally the order decays into deterministic chaos as a fractal attractor of complex systems.

In a more mathematical way, the microscopic view of a complex system is described by the evolution equation of a state vector where each component depends on space and time and where the components may denote the velocity components of a fluid, its temperature field, or in the case of chemical reactions, concentrations of chemicals. The slaving principle of synergetics [2] allows us to eliminate the degrees of freedom which refer to the stable modes. In the leading approximation the evolution equation can be transformed into a specific form for the *nonlinearity* which applies to those systems where a competition between patterns occurs. The amplitudes of the leading terms of unstable modes are called *order parameters*. Their evolution equation describes the emergence of macroscopic patterns. The final patterns (*'attractors'*) are reached by a transition which can be understood as a kind of symmetry breaking.

In the framework of complex systems the emergence of life is not contingent, but necessary and lawful in the sense of synergetics. Only the conditions for the emergence of life (for instance on the planet Earth) may be contingent in the universe. Open ("dissipative") physical and chemical systems lose their structure when the input of energy and matter is stopped or changed (e.g., laser, BZ-reaction). Organismic systems (like cells) are able to conserve much of their structure at least for a relatively long time. On the other hand, they need energy and matter in a certain interval of time to keep their structure more or less far from thermal equilibrium. Thus, biological systems combine conservative and dissipative structures which are determined and reproducible by DNA-replication. It is an open question of prebiotic evolution how the phase transition from conservative and dissipative to *DNA-reproducible structures* had been arranged. In any case we have complex systems the development of which can be explained by the evolution of (macroscopic) order parameters caused by nonlinear (microscopic) interactions of molecules, cells, etc., in phase transitions near to or far from thermal equilibrium. Forms of biological systems (plants, animals, etc.) are described by order parameters. Spencer's idea that the evolution of life is characterized by increasing complexity which can be made precise in the context of complex systems. It is well known that Turing analyzed a mathematical model of organisms represented as complex cellular systems. The evolution of the order parameter corresponds to the aggregation forms during the phase transition of the macroscopic organism. The mature multicellular body can be interpreted as the 'goal' or (better) 'attractor' of organic growth.

Even the ecological growth of biological populations may be simulated using the concepts of synergetics. Ecological systems are complex dissipative systems of plants or animals with mutual nonlinear metabolic interactions with each other and with their environment. The symbiosis of two populations with their source of nutrition can be described by three coupled differential equations which were already used by Edward Lorenz to describe the development of weather in meteorology. In the 19th century the Italian mathematicians Lotka und Volterra described the development of two populations in ecological competition. The nonlinear interactions of the two complex populations are determined by two coupled differential equations of prey and predator species. The evolution of the coupled systems have stationary points of equilibrium. The attractors of evolution are periodic oscillations (limit cycles).

3. Synergetics and Complex Computational Systems

Perhaps the most speculative interdisciplinary application of complex systems is the human brain as a multi-cellular system. The emergence of mental states (for instance pattern recognition, feelings, thoughts) is explained by the evolution of (macroscopic) order parameters of cerebral assemblies which are caused by nonlinear (microscopic) interactions of neural cells in learning strategies far from thermal equilibrium. After *conservative* and *dissipative self-organization* of atomic and molecular systems and after *DNA-codified self-organization* of cellular systems, *learning strategies* are understood as neural self-organization of the brain which are not directed by the DNA code. Cell assemblies with mental states are interpreted as attractors (fixed points, periodic, quasi-periodic, or chaotic) of phase transitions. If the brain is regarded as a complex system of neural cells, then its dynamics is assumed to be described by the nonlinear mathematics of neural networks. Pattern recognition, for instance, is interpreted as a kind of phase transition by formal analogy with the evolution equations which are used for pattern emergence in physics, chemistry, and biology.

We get an interdisciplinary research program that should allow us to explain neurocomputational self-organization as a natural consequence of physical, chemical, and neurobiological evolution by common principles. As in the case of pattern formation, a specific pattern of recognition (for instance a prototype face) is described by order parameters to which a specific set of features belongs. Once some of the features which belong to the order parameter are given (for instance a part of a face), the order parameter will complement these with the other features so that the whole system acts as an associative memory (for instance the reconstruction of a stored prototype face from an initially given part of that face). According to the *slaving principle* the features of a recognized pattern correspond to the enslaved subsystems during pattern formation.

The development of neural and synergetic computers is described and contrasted with Turing machines and knowledged-based systems. In *synergetic computers*, the order parameter equations allow a new kind of (non-Hebbian) learning, namely a strategy to minimize the number of synapses. In contrast to neurocomputers of the spin-glass type (for instance Hopfield systems), the neurons are not threshold elements but rather perform simple algebraic manipulations like multiplication and addition. Beside deterministic homogeneous Hopfield networks there are so-called Boltzmann machines with a stochastic network architecture of non-deterministic processor elements and a distributed knowledge representation which is described mathematically by an energy function. While Hopfield systems use a Hebbian learning strategy, Boltzmann machines favour a backpropagation strategy (Widrow-Hoff rule) with hidden neurons in a many-layered network.

In general it is the aim of a learning algorithm to diminish the information-theoretic measure of the discrepancy between the brain's internal model of the world and the real environment via self-organization. The recent revival of interest in the field of neural networks is mainly inspired by the successful technical applications of statistical mechanics and nonlinear dynamics to solid state physics, spin glass physics, chemical parallel computers, optical parallel computers, and -- in the case of synergetic computers -- to laser systems. Other reasons are the recent development of computing resources and the level of technology which make a computational treatment of nonlinear systems more and more feasible. Besides 'artificial intelligence' (AI) as a classical discipline of computer science 'artifical life' (AL) is a new growing field of research in the framework of the sciences of complexity. Natural life on earth is organized on the molecular, cellular, organism, and population-ecosystem level. Artificial life aims at modeling tools powerful enough to capture the key concepts of living systems on these levels of increasing complexity.

John von Neumann's concept of cellular automata gave the first hints of mathematical models of living organisms conceived as self-reproducing networks of cells. The state space is a homogeneous lattice which is divided into equal cells like a chess board. An elementary cellular automaton is a cell which can have different states, for instance "occupied" (by a mark), "free", or "colored". An aggregation of elementary automata is called a composite automaton or configuration. Each automaton is characterized by its environmment, i.e., the neighboring cells. The dynamics of the automata is determined by synchronous transformation rules. Von Neumann proved that the typical feature of living systems, their tendency to reproduce themselves, can be simulated by an automaton with 200 000 cells (in the plane), where each cell has 29 possible states and the four orthogonal neighboring cells as environment. Although this idea is justified by a mathematically precise proof, it is hard to realize by technical computers. The reason is von Neumann's requirement that the self-reproducing structure must be a universal computer which has the complexity degree of a universal Turing machine. Obviously, self-reproducing molecules of prebiotic evolution were hardly capable of universal constructions. If one drops the requirement of universality, very simple cellular automata can be designed that can reproduce themselves.

In general, *cellular automata* have turned out to be *discrete models of complex systems with nonlinear differential equations* describing their evolution dynamics. The time evolution of their rules, characterizing the dynamics of cellular automata, produces very different cellular patterns, starting from random initial conditions. Computer experiments give rise to the following classes of attractors the cellular patterns of evolution are aiming at. After a few steps, systems of class 1 reach a homogeneous state of equilibrium independently of the initial conditions. This final state of equilibrium is visualized by a totally white plane and corresponds to a fixed point as attractor. Systems of class 2 , after a few time steps, show a constant or periodic pattern of evolution which is relatively independent of the initial conditions. Specific positions of the pattern may depend on the initial conditions, but not the global pattern structur itself. Systems of class 3 evolve toward chaotic states as final attractors without any global periodicity. These chaotic patterns depend sensitively on the initial conditions, and show self-similar behavior with fractal dimension. Systems of class 4 produce highly complex structures with locally propagating forms. Systems of class 3 and 4 are sensitive to tiny fluctuations that can effect a global change of order (the "butterfly effect"). Thus, in these cases, the process of evolution cannot be forecast in the long run.

Obviously, these four classes of cellular automata model the attractor behavior of nonlinear complex systems which is well known from self-organizing processes. They can easily be related to our concept of *synergetics with order and control parameters*. Order parameters correspond to macroscopic spatio-temporal properties of cellular patterns such as fixed point, periodic, complex or chaotic attractors. In the synergetic approach, stable and instable collective motions ("modes") can be distinguished close to critical points of instability. This instability is caused by a change of a control parameter and leads to new macroscopic spatio-temporal patterns. In cellular automata, stable and unstable motions can be characterized by transition rules unchanging and changing previous states. Thus, a control parameter of cellular automata must refer to their transition rules and the transient length of their patterns. It must correspond to the intuitive idea that complex systems lie somewhere in the continuum between order (like ice crystals and carbon Buckminsterfullerene) and chaos (like molecules in a gas): Organisms and brains are complex, but neither wholly ordered nor wholly disordered.

The *complexity of discrete cellular automata* can also be related to *physical measures of complexity* [3]. The physical depth of a macroscopic state of a system is equal to the amount of information required to specify the trajectory that the system has followed to its present state. Applied to closed or Hamiltonian systems, the thermodynamical depth of a state is proportional to the difference between the coarse-grained entropy (the state's thermodynamic entropy) and the fine-grained entropy (corresponding to the trajectory that the system has followed to the state). By analogy with the complexity of cellular automata, complex systems with large *thermodynamical depth* lie between completely ordered and irregular systems with vanishing depth. The reason is that systems in a highly ordered and regular state such as salt crystals have vanishing entropies. Further on, systems at thermodynamic equilibrium are not deep ('complex'), because the coarse- and fine-grained entropies of states with uniform distribution equals and, thus, their difference (i.e., the thermodynamical depth) must be zero. On the other hand, the DNA of a living creature has a large depth, because the trajectory whereby the creature developed since the prebiotic evolution requires a great deal of information to specify. In general, the genetic complexity of an organism is proportional to the amount of genetic information tried out and discarded by the process of natural selection on the ancestors of the organism. If one could map out the evolutionary path by which organisms came into being, then one could order organisms in a hierarchy of complexity degrees with differing thermodynamic depth. Thermodynamically deep ('complex') systems are far from thermal equilibrium.

The nonlinear mathematics of complex systems can only be mastered by the increasing of computer technologies. What is the relation between the amount of computational efforts and physical depth of computer capacities? *The computational complexity* of a number is defined as the length of the shortest computer program that produces that number as output. Any information can be represented as a sequence of symbols (e.g., binary string of 0s and 1s). For example, a binary string of 10^6 0s is computationally very simple, because its program of computation is already given by the device 10^6. By contrast, a random string of 10^6 0s and 1s obtained by flipping a coin cannot be compressed to a similarly concise description. Therefore, random and incompressible strings are computationally very complex. If a computer is considered as a physical system, then the computational complexity of a number is proportional to the minimum amount of information required to specify a trajectory that the computer could follow in calculating that number. In general, the computational complexity of a problem is porportional to the thermodynamic depth of the most efficient machine that solves that problem.

Are there limitations of computational models simulating complex natural systems? Some logical difficulties are related to *Gödel's theorem of undecidability*. Gödel's theorem establishes that there is no systematical way to distinguish halting from non-halting computer programs. Undecidability thus defies all attempts to derive the most efficient algorithm in a finite number of steps. Consequently, no algorithm can distinguish computationally random and incompressible strings from simple ones. The minimal programs needed to guarantee the optimum thermodynamic efficiency cannot be found systematically. But these remarks should not be misunderstood. There are only no decision procedures for absolute measures of complexity. But, there are relative measures with practical efficiency.

It is, for example, not difficult to suggest a systematic method of finding the shortest program that can generate a certain string of symbols within a pre-determined finite number of steps. Relative to that number, complexity compression is not possible. Furthon, the halting problem does not forbid the absolutely best solution being found after a finite number of steps (for instance, by a lucky guess). It is only certain that we will never know how lucky we were. Thus, cosmic evolution might have delivered the best of all possible worlds in the sense of Leibniz. But Gödel's theorem implies that the information required to attain maximum efficiency can be secured only through an infinitely long computation. Complex systems with the *complexity degree of an universal Turing machine* could only be simulated by another universal Turing machine. In that case, the efforts of computational simulation could only be compressed with a (not essential) linear factor. Consequently, partial simulations of complex systems such as our brain are not excluded, but they may need practically incompressible efforts of computation [4].

4. Synergetics and Complex Socio-Economical Systems

Self-reinforcing mechanics with positive feedbacks are typical features of nonlinear complex systems in economics and societies. If a product has some accidental competitive advantages on the market, then the market leader will dominate in the long run and even enlarge its advantage without being necessarily the better product. Many examples of modern high-tech industries show that competing products may have approximately equal market shares in the beginning. But tiny fluctuations which increase the market share of a particular product decide its final success. Sometimes the final market leader may even have less quality from a technical point of view. These effects cannot be explained in the framework of traditional linear dynamics. But they are well known in nonlinear systems.

Within an economy there are several markets with particular dynamics (e.g., a self-organizing stock market with crashes and chaos). They are influenced by cycles, for instance, the annual solar cycle determining the agricultural, tourist, or fuel market. The pig cycle and building cycle are further well known economic examples. Thus endogenous nonlinear systems with impressed waves of exogenous forces are realistic models of economies. The impression given is of a disturbed chaotic attractor or a kind of super-chaos. It is this erratic character of economic events which causes great problems for economic agents who have to make decisions that depend on an unpredictable future.

Historically, the Great Depression of the 1930s inspired economic models of business cycles. However, the first models (for instance Hansen-Samuelson and Lundberg-Metzler) were linear and hence required exogenous shocks to explain their irregularity. The standard econometric methodology has argued in this tradition, although an intrinsic analysis of cycles has been possible since the mathematical discovery of strange attractors. The traditional linear models of the 1930s can easily be reformulated in the framework of nonlinear systems. There are remarkable examples of chaotic ('strange') attractors in economies. Although each trajectory is exactly determined by the evolution equations, it cannot be calculated and predicted in the long run. Tiny deviations of the initial conditions may dramatically change the path of trajectories in the sense of the butterfly effect.

The theory of complex systems can help to design a *global phase portrait of an economic dynamics*. But experience and intuition are sometimes more helpful than scientific knowledge for finding the local equilibria of economic welfare. Politicians themselves must have high sensitivity to deal with highly sensitive complex systems. Obviously, a complex market cannot be commanded such as an army. The Laplacean spirit of a planned market has been falsified by recent historical experiences. On the other hand, historical experience shows that a self-organizing market does not deliver human welfare automatically such as molecular self-organization producing a nice crystal. We need appropriate social conditions ('control parameters') of economic self-organization, in order to serve the people and to produce welfare.

Human society of consumption and production must be embedded in the complex equilibria and cycles of nature (e.g., by industrial recycling). We all know that short-term advantages (e.g., profits of industries, welfare of consumers) can cause a global collapse of our living conditions. We all know the dramatical conflict which is provoked by the exponential growth of human population, industries, and agricultural activities and the increasing stress of climate. Human work is based on natural energies such as sun, wind, water, fossil and nuclear energies. The theory of complex systems may help us to find an appropriate strategy of energy, climate and human welfare with respect to the cycles and equilibria in the economical-ecological system.

In the *framework of synergetics* the behaviour of human populations is explained by the evolution of (macroscopic) *order parameters* which is caused by *nonlinear* (microscopic) interactions of humans or human subgroups (states, institutions, etc.). *Social or economic order* is interpreted by *attractors of phase transitions*. For example, consider the growth of urban regions. From a microscopic point of view the evolution of populations in single urban regions is mathematically described by coupled differential equations with terms and functions referring to the capacity, economic production, etc., of each region. The macroscopic development of the whole system is illustrated by computer-assisted graphics with changing centers of industrialization, recreation, etc., caused by nonlinear interactions of individual urban regions (for instance advantages and disadvantages of far and near connections of transport, communication, etc.). An essential result of the synergetic model is that urban development cannot only be explained by the free will of single persons. Although people of local regions are acting with their individual intentions, plans, etc., the tendency of the global development is the result of nonlinear interactions.

Computers and informational systems have become crucial techniques in sociocultural development, evolving in a quasi-evolutionary process. The replicators of this process are any of the information patterns that make up a culture and spread with variation from human to human. As humans, unlike molecules or primitive organisms, have their own intentionality, the spreading process of information patterns is realized not via mechanical imitation but via communication.The capability to manage the complexity of modern societies depends decisively on an effective communication network. Like the neural nets of biological brains, this network determines the learning capability that can help mankind to survive. In the framework of complex systems, we have to model the dynamics of information technologies spreading in their economic and cultural environment. Thus, we speak of informational and computational ecologies. There are actually realized examples, like those used in airline reservation, bank connections, or research laboratories, which include networks containing many different kind of computers.

The *growth of informational and computational ecosystems* is connected with a fundamental change of society characterized by a switch from traditional industries handling goods to knowledge industries for information and an economy of information services. The production, distribution, and administration of information have become the main acitivites of modern knowledge based societies. Thus, the interface between humans and information systems must be continously improved in order to realize the ideal of worldwide human communication. Human means of expression like speech, gestures, or handwriting should be understood immediately by computation and information systems. The 'whole-person paradigm' and the 'human-machine network' are highlights in the future world of communication.

5.   Synergetics and Complex Science and Technology

The development of our society at the turn of the century depends essentially on the growth of science and technology. Progress in science seems to be governed by the complex dynamics of scientific ideas and research groups which are embedded in the complex network of human civilization. Common topics of research attract the interest and capacity of researchers for greater or lesser periods of time. These 'attractors' of research seem to dominate the activities of scientists like the attractors and vortices in fluid dynamics. When states of research become unstable, research groups may split up into subgroups following particular paths of research which may end with problem solutions or may bifurcate again, and so forth. *Progress in science* seems to be realized by *phase transitions* in a *bifurcation tree with increasing complexity*. Sometimes scientific problems are well-defined and lead to clear problem solutions. But there are also "strange" and "diffuse" states like the strange attractors of chaos theory. How can we manage the complex process? From a microscopic point of view, local interactions of scientists with their individual abilities, ideas and interests determine the complex dynamics of research. From a macroscopic point of view, the global dynamics of research fields is determined by attractors (*'order parameters'*) of research. Further on, there is a strong interaction of research and its environment, i.e. society, politics, economy. Are there appropriate models of scientific growth?

Consider a system with an enumerable set of fields $i = 1,2,3,...$ (e.g., subdisciplines of physics), each of which is characterized by a number of occupying elements N_1, N_2, $N_3...$ (e.g., physicists working in the particular subdiscipline). Elementary processes of self-reproduction, decline, exchange, and input from external sources or spontaneous generation have to be modeled. By a generalization of Eigen's equation of prebiotic evolution, the evolutionary processes of selfreproduction, decline, and field mobility can be scientometrically interpreted in the following way. Selfreproduction means that young scientists join the field of research they want to start working in. Their choice is influenced by education processes, social needs, individual interest, scientific school, etc. Decline means that scientists are active in science for a limited number of years. The scientists may leave the scientific system for different reasons (e.g., the reason of age). Field mobility means the exchange process ('migration') of scientists between research fields. For higher-order effects and cross-catalytic phenomena, the evolution equation must be generalized to include nonlinear effects of growth, decline, and transition rates.

Actually, scientific evolution is a stochastic process. When, e.g., only a few pioneers are working in the initial phase of a new field, stochastic fluctuations are typical. A high correlation between field mobility processes and the emergence of new fields has been found. The stochastic dynamics of the probability density of the scientific fields is modeled by a master equation, where the transition operator is defined by the transition probabilities of field selfreproduction, field decline, field mobility and spontaneous generation. The stochastic model provides the basis for several computer-assisted simulations of scientific growth processes. The corresponding deterministic curves as average over a large number of identical stochastic systems are considered for trend analysis, too. As a result, the general S-shaped growth law of scientific communities in subdisciplines with a delayed initial phase, a rapid growth phase, and a saturation phase has been established in several simulations. The S-shaped nonlinear logistic map gives rise to a variety of complex dynamic behaviors such as fixed points, oscilllations, and deterministic chaos, if the appropriate control parameters increase beyond critical threshold values. Obviously, the stochastic as well as the deterministic models reflect some *typical properties of scientific growth*. Such effects are structural differentiation, deletion, creation, extension of new fields with delay, disappearance, rapid growth, overshooting fashions, and retrogration. The computer-assisted graphic simulations of these dynamical effects can be characterized by appropriate order parameters which are testable on the basis of scientometric dates. Possible scenarios under varying conditions can be simulated, in order to specify the landmarks and the scope of future developments.

6. Outlook on the Future of Civilization: Synergetics, Complexity, and Responsibility

The principles of synergetics and complex systems suggest that the natural, economic and technical world is *nonlinear and complex*. This essential result has important consequences for our present and future behavior. What can we know about its future? What should we do? *Forecasting* in a linear and conservative world without friction and irreversibility would be perfect. We only need to know the exact initial conditions and equations of a running process, in order to predict future by solving equations for future events. But patterns and relationships in economics,business, and society change sometimes dramatically. In addition to the natural sciences, people's actions which are observed in the social sciences can and do influence future events. A forecast can, therefore, become a self-fulfilling or self-defeating prophecy that changes established patterns or relationships of the past. Is forecasting nothing more than crystal-balling?

Forecasting the future of technological trends and markets, the profitability of new products or services, and the involved development of employment and unemployment is one of the most difficult, but also most necessary tasks of managers and politicians. Their decisions depend on a large number of technological, economic, competitive, social, and political factors. In general, the mathematical methods of forecasting followed the line of linear thinking. On the other hand, the increasing capability of modern computers encouraged researchers to analyze nonlinear problems. In the mid-1950s meteorologists preferred statistical methods of forecasting based on the concept of linear regression. This development was supported by Norbert Wiener's successful predicting of stationary random processes. Edward Lorenz was sceptical about the idea of statistical forecasting and decided to test its validity experimentally against a nonlinear dynamical model.

Weather and climate is an example of an open system with energy dissipation. The state of such a system is modeled by a point in a phase space, the behavior of the system by a phase trajectory. After some transient process a trajectory reaches an attracting set ('attractor') which may be a stable singular point of the system, a periodic oscillation called a limit cycle or a strange attractor. If one wants to predict the behavior of a system containing a stable singular point or a limit cycle, the divergence of nearby trajectories does not appear to be growing and may even diminish. In this case, a whole class of initial conditions will be able to reach the steady state and the corresponding systems are predictable.

When the corresponding *nonlinear differential equations* are *digitalized for computation*, sensitivity for initial data translates into a finite average information flow rate of digit strings which is measured by the Liapunov exponent. Any error change of digitalized initial data of a digitalized chaotic dynamics leads to an exponentially increasing computational time of future data. Thus, *computational complexity* in the sense of *computational time* becomes a *practical limitation of long-time prediction* in chaotic and complex systems, although their future trajectories may be determined in principle. It is an interesting practical question whether concepts of parallelism and connectionism which are assumed to be typical for neural networks of the brain can be used to speed up the computation of chaotic and complex systems.

Nevertheless, with the interdisciplinary application of complex systems, we get general insights in the predictable horizons of oscillatory chemical reactions, fluctuations of species, populations, fluid turbulence, or economic processes. The emergence of sunspots, for instance, which was formerly analyzed by statistical methods of time-series is by no means a random activity. It can be modeled by a nonlinear chaotic system with several characteristic periods and a strange attractor only allowing bounded forecasts of the variations. In nonlinear models of public opinion formation, for instance, we may distinguish a predictable stable state before the public voting ('*bifurcation*') when none of two possible opinions are preferred, the short interval of bifurcation when tiny unpredictable fluctuations may influence abrupt changes, and the transition to a stable majority. The situation reminds us of growing air bubbles in turbulently boiling water: When a bubble has become big enough, its steady growth on its way upward is predictable. But its origin and early growth is a question of random fluctuation. Obviously, nonlinear modeling explains the difficulties of the modern Pythias and Sibyls of demoscopy.

The *complexity of sociocultural evolution* could allow several attractors. They cannot be forecast or determined by human decisions, but they may be influenced by conditions and constraints humans are able to achieve. What is the chance of *human freedom in a world of high complexity*? What is the degree of individual responsibility in a complex world of collective effects with high nonlinearity? As the ecological, economic, and political problems of mankind have become *global, complex,* and *nonlinear*, the traditional concept of individual *responsibility* is questionable. We need new models of collective behavior depending on the different degrees of our individual faculties and insights. Individual freedom of decision is not abolished, but restricted by collective effects of complex systems in nature and society which cannot be forecast or controlled in the long run. Thus, it is not enough to have good individual intentions. We have to consider their nonlinear effects. *Global dynamical phase portraits* deliver *possible scenarios* under certain circumstances. They may help to achieve the appropriate conditions for fostering desired developments and preventing evil ones.

The *ethical consequences* strongly depend on our knowledge about complex nonlinear dynamics in nature and society. They invite a demand for *sensitivity* in dealing with highly sensitive complex systems in nature and society. We should neither overact nor retire, because overaction as well as retirement can push the system from one chaotic state to another. *We should be both cautious and courageous, according to the conditions of nonlinearity and complexity in evolution.* In politics we should be aware that any kind of mono-causality may lead to dogmatism, intolerance, and fanaticism. But there is no algorithm to generate the best of all possible worlds. Dynamical models of urban developments, global ecologies or information networks only deliver possible scenarios with different attractors. It is a challenge for us to evaluate which attractor we should prefer ethically and help to realize by achievement of the appropriate conditions.

References:

[1]   K. Mainzer, *Thinking in Complexity: The Complex Dynamics of Matter, Mind, and Mankind*, Springer, Berlin, Heidelberg, New Yorks 1994, 3rd enlarged edition 1997 (Japanese translation of the 2nd edition 1996: Tokyo 1997, Chinese Translation: Beijing 1997).

[2]   H. Haken, *Synergetics*, Berlin, Heidelberg, New York 1977, 3rd enlarged edition 1983, 195; H. Haken/A. Mikhailov (eds.), *Interdisciplinary Approaches to Nonlinear Complex Systems*, Springer, Berlin, Heidelberg, New York 1993.

[3]   S. Lloyd, H. Pagels, *Complexity as Thermodynamic Depth*, Annals of Physics 188 1988, 186-213.

[4]   K. Mainzer, *Gehirn, Computer, Komplexität*, Springer, Berlin, Heidelberg, New York 1997.

The set of boundary points as a complex system

Masaya Yamaguti

Dept. of Applied Math. and Informatics, Ryukoku Univ., Seta, Ohtsu 520-21, JAPAN

Abstract

My proposal is to study boundary point set as a comlex system. As an example,
the boundary of Mandelbrot set is very complex whose Haussdorff dimension is proved
2 by Prof. Shishikura 1993. Next, I will explain the study of Logic by George Boole
who proved the principle of Contradiction by an assumption. This shows that some
border of one set is usually very complicated. Then, I will discuss many problems
in Engineering Economics, Politics and Psychiatry.

Key words: Complexity, Mathematical logic, Border set

1. One example of a set of boundary points

Let me show the boundary of a Mandelbrot set. Fig. 1 is a sequence of figures of
a Mandelbrot set by successive magnifications. You can see easily that each small
part has a great complexity. After several maginification we found total original
figure again.
Prof. M. Shishikura has proved in 1993 that this set of boundary points has Hausdorff
dimension 2. I would like to say that this kind of boundary sets appears very often
in every field of Science and Technology and also many difficult border problem in
Polytics and Economy.
Finally I can point out the same problem in Psychiatry. Namely, the bondary of Self
and Others.
Before discussing these points, I will mention about the work of one English Mathe-
matician G. B. and his life.

2. G. Boole, his life and work

George Boole was born in 1815 as the eldest son of John Boole at Lincolnshire in
England. George learnd English language and grammar from his father. But he was
very talented in Language. Already when he was 14 years old, he translated a greek
poem in English. His translation "Ode to Spring" was published in one column of one
local news paper(Lincoln Herald). The people of Lincolnshire were surprised by the
fact that this poem had been beautifully translated into English by one teen-ager.
Until this age, George had completely finish his study of Greek, Latin and French
by himself. Also, he had finish by himself his study of elemantary geometry at 11
years old. His father was very poor shoemaker, therefore he could not get some fund
to go to University. But, he worked as a teacher of private school. Beside his
daily works as a teacher, he was volunteer for 3 social works; namely

(1)Mechanic Institute(promotion of sciences for people)
(2)Female Penitent's Home
(3)Early closing Association
He studied higher mathematics in the Library of Mechanic Institute.

3. His Mathematics

His famous book entitled "An investigation of laws of thought" was published in
1854. Let me explain a little about his idea.
The laws of thought is to investigate the fundamental laws of those operations of
the mind by which reasoning is perfomed. He wanted to establish the science of
logic. He considered, if the symbol x represents the class of blue objects and the
symbol y, the class of round objefts, then the compound symbol $x y$ to represents
the class of objects that are simultaneously blue and round. His symbol obeyed the
commutative law $x y - y x$. Many of the other laws of ordinary multiplication of
numbers are also obeyed.
In the case where every member of the class x is also a member of the class y,
the law of combination of classes gives $x y - x$, and in the special case $x - y$,
an allowable possibility, this equation becomes $x x - x$. By analogy with the usual
multiplication, this is written $x^2 - x$, an equation which is not in general satisfied by numbers, and this equation marks the point where Boolean algebra parts
company with traditional numerical algebra.
Again, if x and y are mutually exclusive classes(that is to say, there is no
object which is simultaneously a member of x and y), Boole defines $x + y$ to represent the class of all objects which belong to class x or class y. Thus, if x
is the class of all men and y is the class of all women, $x + y$ is the class of all
people. One immediately sees that $y + x$, and the other laws resembling those of
addition of numbers are valid. Furthermore, the two operations thus defined are
connected by a law which is $z (x + y) - z x + z y$. (Distributive Law).
Boole made a further important discovery that strengthened this link between
language and mathematics, namely, that there is a perfect analogy between calculus
of classes and culculus of simple propositions.
Thus simple propositions such as "snow is black" or "London is a city" could be
manipulated analyzed in combination is precisely the same way as classes could.
By analogy with numerical algebra, he assigned the symbol 0 to the empty class
and symbol 1 to the universal class.
This symbol 0 and 1 work exactly same manner as in numerical algebra. That is to
say $0 x - 0$ and $1 x - x$ for x any class of objects.
This came from the religious belief that 1 is God, God is the universe.
Here, I have to explain his religion.
He was baptized, like the majority Englishman at the time, in the Church of England
and his childhood religous observance included attendance at church services, and
Sunday school, and frequent reading Bible. But he cannot believe Trinity, he de-
veloped conscientious objection against the signing of the Thirty Nine Articles.
He was near Unitarian.
Now, it is a time to show his proof of "Principle of Contradiction"(Aristotle's
axiom). We already shown the equality: $x^2 - x$ by his discussion. From this equality,
He could prove the Principle of Contradiction.
$x^2 + x \Rightarrow x - x^2 - 0$
$0 - x - x^2 - x 1 - x \ x - x (1 - x) - 0$
$1 - x$; class of all object except x

Consequently

1-x-non x

x•(non x)-0-empty

This is exactly "Principle of Contradiction".

But, we shall not to forget the following remark in his own book "An Investigation of Law of Thought": The law which is exprsses is practically examplified in language.

To say "good, good", in relation to any subject, though a cumbrous and useless pleonasm, is the same as to say "good". Thus "good, good" men, is the same as to say "good" men.

Such repetitions of words are indeed sometimes employed to heighten a quality or strenghten an affirmation.

But this effect is secondary and conventional: it is not founded in intrinsic relation of language and thought.

4. Engineering

First, I will talk about Engineering. Usually one take mechanical system as a counter exmple of complex system. But I don't agree to this opinion. Because a mechanical system consists of many parts. How about the joining part of each two parts. This is really very comlex. Our modern science avoid this difficulty. We don't know about the phenomenon arising in a joint part between an axis and a wheel which are not connected in a vehicle in Shinkansen.

5. Psychiatry

Here, the boundary is between Self and Others.

Jacques Lacan(1978) used the golden ratio ($\sqrt{5}$-1)/2 as the ideal ratio between Self and the whole(Self and Other). But I am studying the following model Fig. 2 of Self and Others.

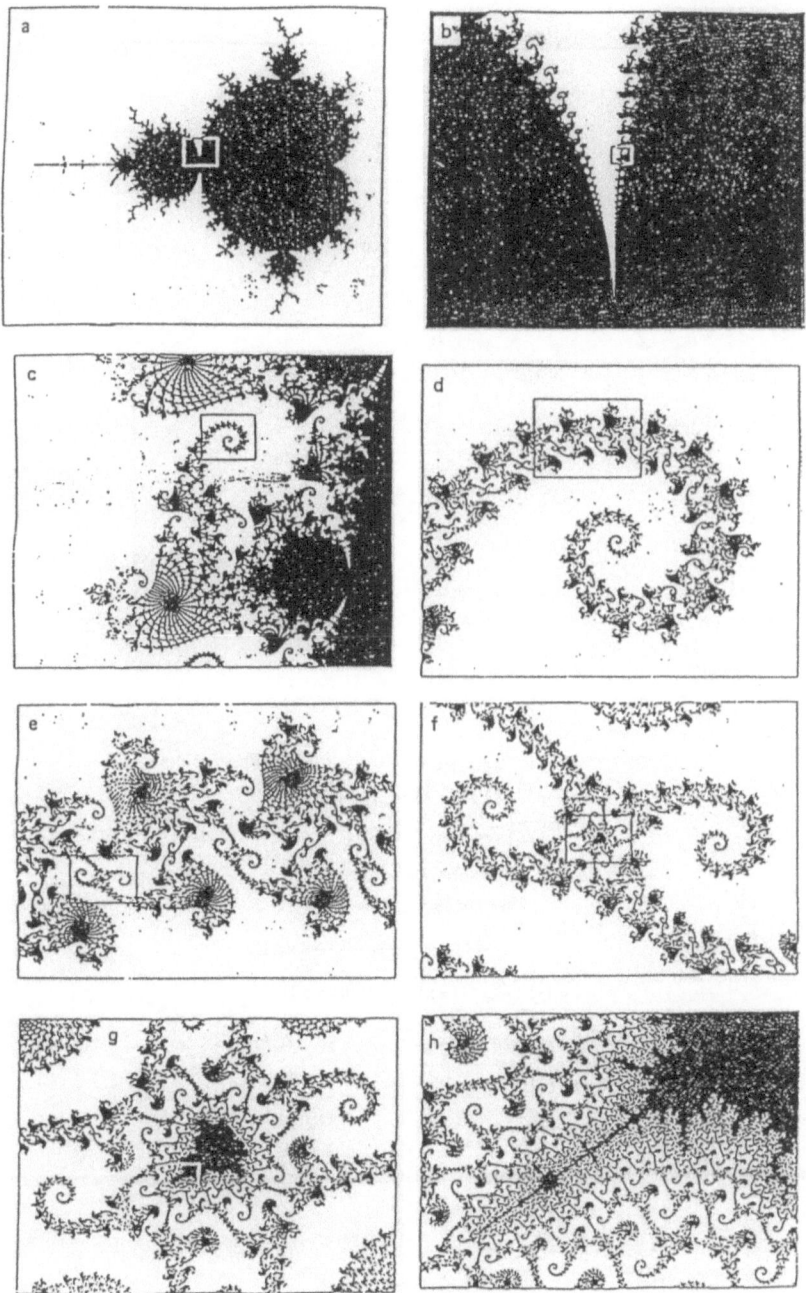

Fig. 1

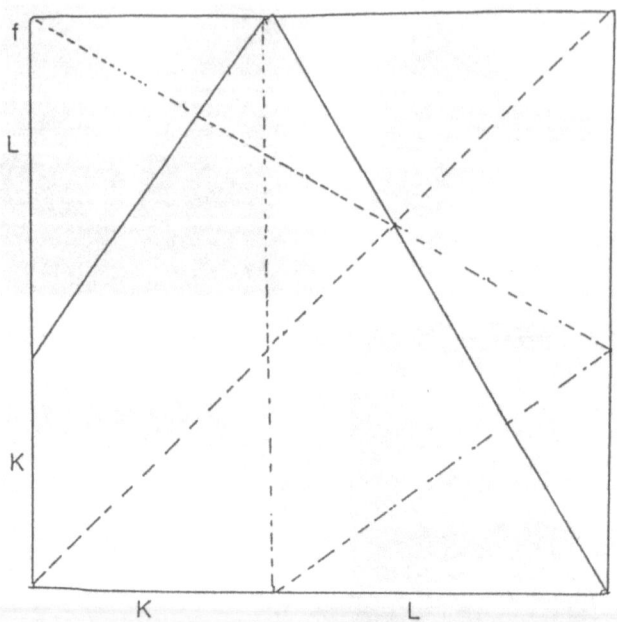

$$x_{n+1} = f(x_n)$$

$$f(x) = \begin{cases} \dfrac{\sqrt{5}+1}{2}x + \dfrac{3-\sqrt{5}}{2} & (0 \le x \le \dfrac{3-\sqrt{5}}{2}) \\[2mm] \dfrac{\sqrt{5}+1}{2}(1-x) & (\dfrac{3-\sqrt{5}}{2} \le x \le 1) \end{cases}$$

$$f(K) = L, \ f(L) = L + K = [0, 1]$$

$$g_n(x) : \begin{cases} 0 \, (0 \le x \le \dfrac{3-\sqrt{5}}{2}) & 1 \to K \\[2mm] \dfrac{\sqrt{5}-1}{2}(x - \dfrac{3-\sqrt{5}}{2}) & (\dfrac{3-\sqrt{5}}{2} \le x \le 1) \end{cases}$$

$$g_1(x) : \dfrac{\sqrt{5}-1}{2}(1-x) + \dfrac{3-\sqrt{5}}{2} \, (0 \le x \le 1)$$

$$(\dfrac{\sqrt{5}+1}{2})(1-x) = x$$

$$(1-x) = (\dfrac{\sqrt{5}-1}{2})x$$

$$1 = (1 + \dfrac{\sqrt{5}-1}{2})x$$

$$= \dfrac{1+\sqrt{5}}{2}x$$

$$x = \dfrac{\sqrt{5}-1}{2}$$

$$1 - (\dfrac{\sqrt{5}-1}{2})x$$

$$x_{n+1} = 1 - ax_n$$

$$x = 1 - ax$$

$$(1+\alpha)x = 1$$

$$x = \dfrac{1}{1 \cdot \int \alpha} = \alpha$$

Fig. 2

Part II Mathematical Systems

On the Complexity of Logic-Dynamics in Brain

Ichiro Tsuda
Applied Mathematics and Complex Systems Research Group,
Department of Mathematics, Graduate School of Science,
Hokkaido University, Sapporo, 060 Japan

Abstract

The complexity theory of interactions among functional levels in brain is developed. In particular, a possible interplay between the logical process and the sensory information processings is highlighted. In this respect, we deal with the transformation between a logical statement and a dynamical system, in relation with dynamics of neurons or neural assemblies. Both internal and external views for the threshold or the switching mechanism of neuron is discussed. An external view is generally represented by dynamical systems, which is assured here by the plausibility of conventional population dynamics. The internal view in the present context is assured by the existence of logical process of macromolecules in the pre- and/or post-synaptic membranes of neuron. Thus the internal view provides logic-based dynamics. There always exist two interpretations for the logic-based dynamics, which lead internal and external dynamics, respectively. The internal dynamics brings about the threshold-related characteristics of neuron, due to its chaotic behaviors, whereas the external one provides a periodic behavior only. With this internal theory, we develop the hermeneutic theory for interfacial dynamics of different levels of neural assemblies, and for the notion of stability of self-description.

Key words: Logic-based dynamics, internal observer, cerebral hermeneutics, chaos, dynamic brain.

1. Introduction

In the recent paper, we have proposed an internal theory for a genesis of biological threshold [1], where "internal" means that material interactions are understood in a constructive way according to causation [2]. One way of the description of the causative relation is to provide explicit descriptions of such interactions according to inference processes with some appropriate logic. Logics are internal ones such as manifesting via behaviors or actions [3], or being constructed from within to represent entity of the world [4].

Assuming the existence of dynamics corresponding to this logical process, one can describe this process in terms of dynamical systems transformed from the logical process. In this respect, we investigated the relation between logic and dynamics in case of the kinase and phosphotase complex which acts for, for instance, morphological changes of ion channel proteins in the post-synaptic membrane of neuron [1]. We briefly review this previous work in §2. In §3, dynamic features of brain are addressed in relation with cerebral Hermeneutics [5, 6], where the basic structure of logic-dynamics is extended to apply to neural assemblies.

2. Logic-Dynamics

We have proposed a new dynamical theory [1] which provides an explicit description concerning a logic which may underlie the dynamic behaviors of complex networks consisting of macro-molecules. This study was motivated by the constructive approaches to neural systems, where at least a formal neuron must be constructed in computer. An indication of formal neuron is provided by the existence of threshold. The proposed mechanisms so far on the threshold phenomena in excitable systems are as follows: The generation of stable manifold of saddle under the appearance of a pair of saddle and node via saddle-node bifurcation [7], the global nonuniformity of vector field under the condition of monostable state [8, 9, 10, 11], and the nonuniform transformation of invariant measure of the patchy and folding map derived from the logic-based dynamics [1].

The assumption of our theory was the presence of logical basis of biological threshold. We took into account, in the model study, ion channel proteins activated or inactivated by the kinase-phosphatase network. This system has a rich complexity such as a complex enzymatic reactions. In spite of this complexity, one can extract the logic underlying these reactions. In our approach, we neglected the details of reactions, rather tried to express the logical sturucture of the reactions as a statement. Thus reaction scheme was converted to an inference process of the statements, and then the inference process was replaced by dynamical system, introducing a discrete time. By dynamical systems analysis, we concluded that the situation of open-and-close of ion channel proteins or their networks expressed by the specific logic can be cause of threshold if they "read" the network behaviors as a self-referential statement according to a continuous logic like the Lukasiewicz logic [12].

The Lukasiewicz logic was first invented as a multi-valued (especially, a three-valued) logic, but extended to a continuous logic in relation with a fuzzy logic [13, 14]. The Lukasiewicz logic is as follows. Let A and B be a statement, and $v(X)$ be a truth value of the statement X.

$$
\begin{align}
v(\neg A) &= 1 - v(A), \tag{1}\\
v(A \wedge B) &= min[v(A), v(B)], \tag{2}\\
v(A \vee B) &= max[v(A), v(B)], \tag{3}\\
v(A \rightarrow B) &= min[1, 1 - v(A) + v(B)], \tag{4}\\
v(A \leftrightarrow B) &= 1 - |v(A) - v(B)|. \tag{5}
\end{align}
$$

Here, $\neg$ denotes *negation*, $\wedge$ *conjunction*, $\vee$ *disjunction*, $\rightarrow$ *imply*, and $\leftrightarrow$ *if and only if*. The truth value is here defined on the unit interval, [0,1].

In case of a diadic logic, only by two operations, *negation* and *conjunction*, any other operations are uniquely derived. Since the truth value is extended to real values on the unit interval, the uniquness of interpretation (the right hand sides) for operations does no longer hold. One can have plural expressions of each right hand side of the above equations, all of which are identical in case of the dyadic truth value. In the above logic, the operation $\vee$ is derived from two operations $\neg$ and $\wedge$, and the operation $\leftrightarrow$ is derived from the operation $\rightarrow$ but not derived from $\neg$ and $\wedge$. The operation $\rightarrow$ derived from $\neg$ and $\wedge$ has a different form from the above. Therefore, a general question in case of a continuous truth value is how one defines *imply*. Actually, several interpretations have been proposed [12, 13]. Among others, we chose the above form, since we believe it would be natural in biological operations.

We introduced a dynamical system for the inference process concerning the truth value of statements describing the occurrence of events, as in the Grim's framework [14]. Let us consider a system S. Suppose N events occur in S. We make N statements, each corresponding to one of N events. We are concerned with the determination process of their truth values. Each statement $P_j(e_1, e_2, \cdots, e_N)$ concerning events $\{e_i\}$ is viewed as *premise*. The corresponding *consequence* $C_j(e_j)$ is defined by the statement concerning the event e_j. The truth value of the statement P_j at time n determines the truth value of the statement C_j at time $n + 1$. Thus we obtain n-dimensional dynamical systems (maps):

$$v_{n+1}(C_j(e_j)) = v_n(P_j(e_1, \cdots, e_N)), j = 1, \cdots, N. \tag{6}$$

There are two interpretations satisfying this framework. Let P and C be a *premise* and a *consequence*, respectively.

(1)

$$v_{n+1}(C) = v_{n+1}(P), \tag{7}$$
$$v_{n+1}(P) = v_n(C). \tag{8}$$

Namely, a logic itself is instantaneous, but the process that views the preceding consequence as the premise is the carrier of time step. This time step is associated with *internal measurements*.

(2)

$$v_{n+1}(C) = v_n(P), \tag{9}$$
$$v_n(P) = v_n(C). \tag{10}$$

Namely, a logic itself from P to C is temporal, but the process from the preceding consequence to the premise is instantaneous. Thus, the transformation from P to C is the carrier of time step. In this interpretation, internal measurements are viewed as to be made instantaneously.

These two interpretations are different resolutions of the same dynamical system, $v_{n+1}(C) = v_n(C)$, in principle. Each could, however, provide different dynamics, if the dynamics transformed from the statement possesses a domain of definition divided into several pieces, and if there exists the mechanism of resetting the truth value to some constant value. This condition is satisfied in the self-referential statement of the second order. In our previous paper [1], we showed the difference in the case of the k-p network statements.

One can adopt a statement of interactions between *excitatory* and *inhibitory* neurons, as in the way previously shown. Let $< e >$ be the statement that an excitatory neuron, or its neural assembly is excited, and similarly $< i >$ for an inhibitory neuron, or its neural assembly.

$$
\begin{array}{ll}
Consequence & Premise \\
< e > \Leftarrow < e > \leftrightarrow < e > \wedge < \neg i >, & (11) \\
< i > \Leftarrow < i > \leftrightarrow < e > \wedge < \neg i > . & (12)
\end{array}
$$

Let x and y be the truth value for $< e >$ and $< i >$, respectively. The premise for an excitatory neuron or its assembly leads to $1 - |x - min(x, 1 - y)|$, and for an inhibitory neuron or its assembly $1 - |y - min(x, 1 - y)|$. Hence, the line $x + y = 1$ devides the unit square into two regions, called region I ($x + y < 1$) and region II ($x + y \geq 1$), respectively. Therefore, the equations of motions according to the interpretation (1) are as follows.

In region I,

$$x_{n+1} = 1, \tag{13}$$
$$y_{n+1} = y_n. \tag{14}$$

In region II,

$$x_{n+1} = 2 - x_n - y_n, \tag{15}$$
$$y_{n+1} = 1 - |1 - 2y_n|, \tag{16}$$

where let each logical inference in respective region I and II be separately instantaneous.

In region I, the dynamics brings about a uniform contraction of area, and chaos-driven dynamics appears in region II. Especially in region II, y-variable obeys a tent map which gives rise to chaos. Hence, the resultant dynamics is a composition of contraction and chaos-driven dynamics, which gives rise to nowhere-differentiable attractors [1, 15, 16, 17, 18, 19, 20]. We obtained a sigmoidal function for the invariant measure of x and an inverse sigmoidal function for y. Attractor itself in $x - y$ plane is nowhere differentiable.

On the other hand, if we adopt another interpretation (2), eq.14 should be replaced by $y_{n+1} = 1 - |y_n - x_n|$. Then, the equations of motion are as follows.

In region I,

$$x_{n+1} = 1, \tag{17}$$
$$y_{n+1} = 1 - |y_n - x_n|. \tag{18}$$

In region II,

$$x_{n+1} = 2 - x_n - y_n, \tag{19}$$
$$y_{n+1} = 1 - |1 - 2y_n|. \tag{20}$$

The overall dynamics shows a periodic orbit only. In all simulations through eqs.13-20, the tent map was modified a bit to avoid the effect of finite computations due to dyadic expansions in computer.

Futhermore, it turned out that functional equations which obey similar dynamical rules to the arguments x and y showed threshold-like phenomena. This was investigated for eqs.13-16, introducing the function $Z(u)$. Here, $Z(u)$ represents a *description* of u, and its value denotes a *certainty* of the description. The following assumption was adopted.

$$Z_{false}(x) = Z_{true}(1 - x). \tag{21}$$

Then, we obtained the functional equations.

In region I,

$$f_{n+1}(x_{n+1}, y_{n+1}) \ = \ 1, \tag{22}$$
$$g_{n+1}(x_{n+1}, y_{n+1}) \ = \ g_n(x_n, y_n). \tag{23}$$

In region II,

$$f_{n+1}(x_{n+1}, y_{n+1}) \ = \ 1 - |f_n(x_n, y_n) - g_n(x_n, 1 - y_n)|, \tag{24}$$
$$g_{n+1}(x_{n+1}, y_{n+1}) \ = \ 1 - |g_n(x_n, y_n) - g_n(x_n, 1 - y_n)|. \tag{25}$$

Here, f is the certainty of self-description for the truth value x affected by y through the equations of motions of x and y, $i.e.$, eqs.13-16. The function g is a similar one for the truth value y. Thus the functions f and g have both arguments x and y, respectively. We call the dynamics of x and y a *lower dynamics*, and the dynamics of f and g a *higher dynamics*. Here a lower and a higher dynamics are almost identical, only the difference of which comes from the relation 21.

By a usage of several techniques, one can solve these functional equations. Actually, we obtained the threshold for a region in $x - y$ space where solutions exist. Thus, we obtained the transformation of continuous variables at the lower level to discrete ones at the higher level. In our previous work [1], a formal neuron was justified in such a way by the lower level's activity of the k-p networks. This theory can also be extended to imply the situation that a sigmoidal behavior of a single neuron or a neural assembly may be represented in a discontinuous form with a help of formation of larger networks of self-description type. Thus the theory brings about an aspect on the interactions between different functional levels in the cortex, which will be discussed in the next section.

3. Dynamic Features of Brain

When one views brain as an assembly of interactive neural assemblies, each consisting of interactive neurons, one notices the following dynamic modalities of brain.

1. *Dynamic functional manifestation at lower levels*

A neuron or a neural assembly in sensory cortices can change a role according to the functional manifestation of prefrontal cortex and/or limbic system. The sensory information, on the other hand, affects the latter. Here, we call the sensory cortex a lower level, and also call both the prefrontal cortex and the limbic system a higher level. What we know from experimental data [3, 21] is that the sensory information conveyed by sensory neurons does not directly denote meanings of external stimuli, rather meanings are created inside due to the higher level's activities. On the other hand, causation of neural dynamics differs in these two levels; the higher level works according to *values* and *logics*, whereas the lower level accords to sensory stimuli; the higher level's activities are comparatively low with the lower level.

In complex systems, in general, the dynamic behaviors that the elementary dynamics does not necessarily obey statistically averaged activities have been observed. Among others, chaotic itinerancy [6, 22, 23] has been widely investigated, also in brain dynamics [3, 6, 21, 24, 25, 26, 27, 28, 29, 30]. By the interaction between different causative systems the lower level's activities become complex ones that can be represented by input-driven chaotic itinerancy [3, 21, 24]. The appearance of chaotic itinerancy can be an indication of the interplay between different levels characterized by, for instance, the different time scales of activities [31]. A functional unit can be represented by this kind of dynamic behavior, thus a functional unit can be reorganized according to the temporal activity level. This dynamic reorganization has been observed in experiments as a *dynamic receptive field* [32, 33, 34] and as a *dynamic tuning of orientation specificity* [32].

2. *Neural Hermeneutics*

In my personal view, one of the most essential notions for an understanding of higher functions of brain is Vorverständnis (preunderstanding) or Vorgriff (grasp in advance) in Hermeneutics, in particular, in Heideggar's one [35]. Brain interprets meanings of external stimuli, based on the sensory information. The purpose of the brain in this context is to *understand* the world. According to Heideggar, however, an interpretation is based on an understanding, but *not* vice versa. An interpretation is prescribed as a process of an understanding itself becoming well-formed. Hence, there must be some structure in an understanding, which is named a preunderstanding. Here, the completeness of meaning is assumed that only the things that possess completely unified meaning can be understood [36]. With a basis of a preunderstanding, a circle between the processes of understanding and of interpreting emerges, which is called a hermeneutic circle [35]. A hermeneutic circle is characterized by an eternal motion of understanding, therefore, it does not stop even after an understanding has become well-formed.

A preunderstanding for meanings of the sensory information in the cortex is seemingly given by the higher level in the form of *values* and *logics* which manifest through actions or behaviors. The basic logic in our theory can be viewed as a kind of a preunderstanding. A self-referential statement may be represented by multi-feedback loops seen at the lower level. Functional equations with dynamic variables viewed as a self-description is the reflection of feedback from the higher level to the lower level. The functional equations do not converge, but a rapid convergence of the domain of (the existence of) solutions is obtained. On the other hand, the number of solutions of the functional equations increases exponentially in time due to the chaotic behavior of the arguments, *i.e.* the lower level. This implies the process of a hermeneutic circle continuing even after the self-description converges.

In general, a process of this kind will be seen in the memory process. A typical network model has been proposed and its dynamic behaviors were precisely investigated [6, 24, 25, 26, 37]. Indirect evidence has been observed in several cortical areas [3, 21, 38, 39, 40].

The proposed framework can be applied to an understanding of a complex system such as macromolecules' networks, metabolic networks, neural networks, and so on, where a huge number of descriptions is demanded if one tries to explicitly write down the details of the reactions, and even when a few dominant degrees of freedom which give rise to only a few descriptions are anticipated, they change in time according to the current background activities. In these systems, it is rare that one can extract the essential dynamics in terms of the notion of *states*, rather a constructive way of understanding based on the notion of *process* is essential [2, 20, 41]. If we describe these systems *externally*, then we would obtain *chaotic itinerancy* as the manifestation of internal conflict. This conflict can be observed *internally* in a projection of chaotic itinerancy to some subspace, where usually nowhere differentiability emerges as the inevitable form of the infinite recursion process which an *internal observer* encounters [20, 42]. This process may be viewed as the process of understanding of internal observer. If it is true, a dynamic representation of a hermeneutic circle will be something nowhere-differentiable. Therefore, the next our concern is the stability problem of *description*.

The formulation of self-description in terms of functional equations leads to the notion of *description* (*in*)*stability* [43, 44, 45]. This is also a generalization of pseudo-orbit-tracing properties in dynamical systems. The pseudo-orbit-tracing property indicates the stability of dynamical systems associated with observations, which is related to structural stability [46]. Our functional equations can be reformulated by a similar scheme [1].

Let us express the functional equation as in the form $f_{n+1}(x_{n+1}) = \tilde{F}(f_n(x_n))$, where $x_{n+1} = F(x_n)$. The function $f(x)$ is viewed as a description of dynamical variable x. Let the dynamical map F and the function f be continuous on compact space M and X, respectively. If there exists $\alpha > 0$ such that $F \circ f_{i-1}^{-1} \circ \tilde{F}^{i-1}$ is in the α neighborhood of $\tilde{F}^i$ for any i, then we say that $\{\tilde{F}^i\}_{i=0,1,\cdots}$ is α pseudo-dynamical system. Furthermore, if for some description $g_0 \in X$ there exists $\beta > 0$ such that $F^n \circ g_0^{-1} \circ \tilde{F}^0$ is in the β neighborhood of $\tilde{F}^n$ for any n, then we say that pseudo-dynamical system $\{\tilde{F}^n\}$ is β-traced by g_0. If any α pseudo-dynamical system is β-traced, let us say that (F, M) has a pseudo-dynamical system-tracing property, or a *description stability*.

Our functional equations have a trivial solution $f^*(x_n) = x_n$ for any n. This *fixed point* in functional space means that the description of dynamical system is just the dynamical system itself, thus implying the existence of "autonomous" dynamical system in the sense of description. A slight perturbation for this identity function, however, makes the descriptions deviate far from the original dynamical system F if F is chaotic. Then, the system concerned belongs to the class of *description instability*. In our hermeneutic context, a preunderstanding is prescribed by the basic logic. The first action is given by the initial function f_0. Since the above dynamic structure remains whatever basic logics are chosen, autonomy in the above sense is inevitably unstable with respect to description, thus a hermeneutic circle becomes inevitable due to chaos.

References

[1] I. Tsuda and K. Tadaki, A Logic-Based Dynamical Theory for A Genesis of Biological Threshold, *Biosystems* (in press, 1997).

[2] K. Matsuno, *Protobiology: Physical Basis of Biology*, CRC press, Boca Raton, Florida, 1989.

[3] W. J. Freeman, *Societies of Brains – A Study in The Neuroscience of Love and Hate*, Lawrence Erlbaum Associates, Inc., Hillsdale, 1995.

[4] G. Boole, *An Investigation of The Laws of Thought*, Dover Publications, Inc., New York, 1854.

[5] I. Tsuda, A Hermeneutic Process of The Brain, *Progress of Theoretical Physics, Supplement*, Vol. 79, 1984, pp. 241-259.

[6] I. Tsuda, Chaotic Itinerancy as A Dynamical Basis of Hermeneutics in Brain and Mind, *World Futures*, Vol. 32, 1991, pp. 167-185.

[7] C. Murakami and K. Tomita, Giant Transitory Excitation– A Thermokinetic Model, *Journal of Theoretical Biology*, Vol. 79, 1979, pp. 203-222.

[8] A. L. Hodgkin and A. F. Huxley, A Quantitative Description of Membrane Current and Its Application to Conduction and Excitation in Nerve, *Journal of Physiology*, Vol. 117, 1952, pp. 500-544.

[9] R. FitzHugh, Impulses and Physiological States in Theoretical Models of Nerve Membrane, *Biophysical Journal*, Vol. 1, 1961, pp. 445-466.

[10] R. FitzHugh, Mathematical Models of Excitation and Propagation in Nerve, in *Biological Engineering*, ed. H. P. Schwan, McGraw-Hill, New York, 1969, pp. 1-85.

[11] J. Nagumo, S. Arimoto, and S. Yoshizawa, An Active Pulse Transmission Line Stimulating Nerve Axon, *Proc. Institute of Radio Engineering*, Vol. 50, 1962, pp. 2061-2070.

[12] N. Rescher, *Many-Valued Logic*, McGraw-Hill, New York, 1969.

[13] L. A. Zadeh, Fuzzy Logic and Approximate Reasoning, *Synthese*, Vol. 30, 1975, pp. 407-428.

[14] P. Grim, Self-Reference and Chaos in Fuzzy Logic, *IEEE Transaction on Fuzzy Systems*, Vol. 1, 1993, pp. 237-253.

[15] J. L. Kaplan and J. A. Yorke, Chaotic Behavior of Multidimensional Difference Equations, *Lecture Notes in Mathematics*, Vol. 730, 1979, pp. 204-227.

[16] C. Grebogi, E. Ott, S. Pelikan, and J. A. Yorke, Starange Attractors That Are Not Chaotic, *Physica D*, Vol. 13, 1984, pp. 261-268.

[17] O. E. Rössler, R. Wais, and R. Rössler, Singular-Continuous Weierstrass Function Attractors, In *Proc. 2nd International Conference on Fuzzy Logic and Neural Networks*, Fuzzy Logic Systems Institute, Iizuka, Japan, 1992, pp. 909-912.

[18] O. E. Rössler, J. L. Hudson, C. Knudsen and I. Tsuda, Nowhere-Differentiable Attractors, *International Journal of Intelligent Systems*, Vol. 10, 1995, pp. 15-23.

[19] O. E. Rössler and J. L. Hudson, A "Superfat" Attractor with A Singular-Continuous 2-D Weierstrass Function in A Cross Section, *Zeitschrift für Naturforshung*, Vol. 48a, pp. 673-678.

[20] I. Tsuda, A New Type of Self-Organization Associated with Chaotic Dynamics in Neural Networks, *International Journal of Neural Systems*, Vol. 7, 1996, pp.451-459.

[21] L. Kay, K. Shimoide and W. J. Freeman, Comparison of EEG Time Series from Rat Olfactory System with Model Composed of Nonlinear Coupled Oscillators, *International Journal of Bifurcation and Chaos*, Vol. 5, pp. 849-858.

[22] K. Ikeda, K. Otsuka and K. Matsumoto, Maxwell-Bloch Turbulence, *Progress of Theoretical Physics, Supplement*, Vol.99, 1989, pp. 295-327.

[23] K. Kaneko, Clustering, Coding, Switching, Hierarchical Ordering, and Control in Network of Chaotic Elements, *Physica D*, Vol. 41, 1990, pp. 137-172.

[24] I. Tsuda, Can Stochastic Renewal of Maps Be A Model for Cerebral Cortex? *Physica D*, Vol.75, 1994, pp. 165-178.

[25] I. Tsuda, Chaotic Neural Networks and Thesaurus, *Neurocomputers and Attention* I, eds., A. V. Holden and V. I. Kryukov, Manchester University Press, Manchester, 1991, pp. 405-424.

[26] I. Tsuda, Dynamic Link of Memories—Chaotic Memory Map in Nonequilibrium Neural Networks, *Neural Networks*, Vol. 5, 1992, pp. 313-326.

[27] S. Nara and P. Davis, Chaotic Wandering and Search in A Cycle-Memory Neural Network, *Progress of Theoretical Physics*, Vol. 88, 1992, pp. 845-855.

[28] S. Nara, P. Davis, M. Kawachi and H. Totsuji, Chaotic Memory Dynamics in A Recurrent Neural Networks with Cycle Memories Embedded by Pseudo-Inverse Method, *International Journal of Bifurcation and Chaos*, Vol. 5, 1995, pp. 1205-1212.

[29] W. J. Freeman, Neural Mechanisms Underlying Destabilization of Cortex by Sensory Input, *Physica D*, Vol. 75, 1994, pp. 151-164.

[30] M. Adachi and K. Aihara, Association Dynamics in a Chaotic Neural Network, *Neural Networks*, (in press, 1997).

[31] H. Okuda and I. Tsuda, A Coupled Chaotic System with Different Time Scales: Possible Implications of Observations by Dynamical Systems, *International Journal of Bifurcation and Chaos*, Vol. 4, 1994, pp.1011-1022.

[32] H. Dinse, A Temporal Structure of Cortical Information Processing, *Concepts in Neuroscience*, Vol. 1, 1990, pp. 199-238.

[33] J. J. Eggermont, A. M. H. J. Aertsen, D. J. Hermes and P. I. M. Johannesma, Spectro-Temporal Characterization of Auditory Neurons: Redundant or Necessary? *Hearing Research*, Vol. 5, 1981, pp. 109-121.

[34] M. Tsukada, a private communication for his experimental findings in late 1970s on dynamic receptive field in cat retinal ganglion cells.

[35] M. Heideggar, *Sein und Zeit*, Tübingen, 1927.

[36] H. G. Gadamer, *Philosophical Hermeneutics*, translated and edited by D. E. Linge, University of California Press, 1976.

[37] I. Tsuda, E. Körner and H. Shimizu, Memory Dynamics in Asynchronous Neural Networks, *Progress of Theoretical Physics*, Vol.78, 1987, pp.51-71.

[38] M. Tsukada, Theoretical Model of The Hippocampal-Cortical Memory System Motivated by Physiological Functions in The Hippocampus, *Cybernetics and Systems: An International Journal*, Vol. 25, 1994, pp.189-206.

[39] M. Tsukada, T. Aihara, M. Mizuno, H. Kato, and K. Ito, Temporal Pattern Sensitivity of Long-Term Potentiation in Hippocampal CA1 Neurons, *Biological Cybernetics*, Vol. 70, 1994, pp.495-503.

[40] Y. Miyashita, Inferior Temporal Cortex: Where Visual Perception Meets Memory, *Annual Review of Neuroscience*, Vol. 16, 1993, pp. 245-263.

[41] Y. -P. Gunji, Autonomous Life as The Proof of Incompleteness and Lawvere's Theorem of Fixed Point, *Applied Mathematics and Computation*, Vol. 61, 1994, pp. 231-267.

[42] Y. -P. Gunji, S. Toyoda and M. Migita, Tree and Loop as Moments for Measurement, *Biosystems*, Vol. 38, 1996, pp. 127-133.

[43] I. Tsuda, deus ex machina–Towards the Brain Theory from Chaos Theory (in Japanese), *System and Control: Transaction of Japan Society of System Engineering*, Vol. 31, 1987, pp. 180-188.

[44] I. Tsuda, *Kaosu-teki Noukan (Chaos Viewpoint of Brain)*(in Japanese), Science-sha, Publ., Inc., Tokyo, 1990.

[45] K. Kaneko and I. Tsuda, *Fukuzatsukei no Kaosu-teki Sinario (Chaos Scenarios of Complex Systems)*(in Japanese), Asakura-shoten, Publ., Inc., Tokyo, 1996.

[46] S. Smale, Differentiable Dynamical Systems, *Bulletin of American Mathematical Soceity*, Vol. 73, 1967, pp. 747-817.

Emergence of Recursivity through Isologous Diversification

Kunihiko Kaneko
Department of Pure and Applied Sciences,
University of Tokyo, Komaba, Meguro-ku, Tokyo 153,Japan

Abstract

A scheme of inter-intra dynamics is presented for the study of biological systems. The dynamics consists of internal dynamics of a unit, interaction among the units, and the replication (and death) of units according to their internal states. Applying the dynamics to cell biology, isologous diversification theory is proposed for cell differentiation. Through several simulations and theoretical considerations, the cell differentiation is shown to proceed through the following steps: (1) Up to some number, cells, created by divisions, are almost identical, whose intra-cellular chemical oscillations are synchronized. (2)As the number exceeds some number, the oscillations lose the synchrony, and cells split into groups of different phases of oscillations. (3)Few distinct gropes of cell types are formed, with different (average) chemical compositions and different types of intra-cellular oscillations. (4) The differentiated behavior of states is transmitted by divisions to daughter cells. Recursivity is formed so that the daughter cells keep the identical chemical character. (5) Hierarchical differentiation proceeds, leading to the generation of the rule of differentiation. Relevance of the theory to cell biology is discussed.

Key words:Inter-intra dynamics, cell differentiation, globally coupled map,isologous diversification, recursivity.

1.Introduction

In a biological system it is important to capture the interplay between inter-unit and intra-unit dynamics. Such "inter-intra dynamics" is essential to cell biology, where complex metabolic reaction dynamics in each unit (cell) are affected by the interaction among cells. An ecological system also consists of interacting units with internal dynamics. In neural systems also, the inter-intra dynamics is relevant to the formation of internal images.

In general, there is an important missing factor in conventional dynamical systems approach to modeling biology. A dynamical system approach consists of time, a set of states, an evolution rule, an initial condition of the states, and boundary conditions. It is generally assumed that these four sets themselves are separated from each other: The set of states itself (e.g., the number of variables) is fixed independent of the evolution rule, and the states cannot alter the evolution rule. Initial and boundary conditions are chosen independently of the states and of the evolution rule.

In biological problems, such separation among the four sets may not be valid. In a biological system, the number of variables itself changes with time, through for example, by cell divisions and cell deaths in the case of cell society, and by the appearance and extinction of new species in the case of population dynamics. Another problem lies in the choice of initial conditions or boundary conditions. For a cell to grow repeatedly, the initial condition of its internal state should satisfy some condition so that the evolution rule leads to daughter cells. Thus, the initial conditions of a state, and its evolution are not clearly separated.

In our inter-intra dynamics approach we try to overcome this separation problem of dynamical systems, by allowing for the change in the degrees of freedom and the formation of a unit acting as a "partial system" which selects its initial and boundary conditions. As an example, the author and Yomo have proposed a novel scenario for cell differentiation termed "*isologous diversification theory*" [1, 2, 3]. The cell differentiation and developmental processes involve internal metabolic reactions, which are nonlinear, as well as cell division and death, which lead to change of the degrees of freedom of the system. Thus, the study of cell differentiation is one prototype of our intra-inter dynamics picture for biological systems.

Indeed, there have been some studies for cellular biology adopting dynamical systems approach. The importance of temporal oscillations in cellular dynamics was studied in pioneering work by Goodwin[4] (see also [5]). The coexistence of many attractors in Boolean network dynamics was attributed to diverse cell types by Kauffman [6]. As for the interaction-based approach in cell differentiation, pattern-formation mechanism of the Turing instability is often adopted. The internal dynamics is based either on stable cycles, or on switching type threshold dynamics allowing only for fixed points. Our inter-intra dynamics approach enables us to discuss the developmental process allowing for differentiation, generation of rules, hierarchical differentiation, and the embedding of ensemble information into internal dynamics.

2. Model for Isologous Diversification

We have studied several abstract models for cell differentiation, to show that the differentiation is a general consequence of a system consisting of replicating units with nonlinear dynamics and interaction [1, 2, 3]. This class of models consists of a metabolic or genetic network within each cell, and interaction between cells through competition for nutrition and the diffusion flow of chemicals to media. In each cell there is a set of chemical variables. When a cell is isolated, its chemicals are assumed to show oscillatory behaviors. As for the inter-dynamics, cells are assumed to interact with each other through the media. The cell volume is expanded through the transport of chemicals into it, until the cell divides into two when the volume exceeds a given threshold. The concentrations of chemicals of two divided cells are chosen to be almost identical upon the division.

We have studied several models with the above setups, from which we extract common scenario for differentiation. Before describing the "isologous diversification theory" for differentiation extracted from the simulation results, we note two backgrounds of the theory, dynamical systems one and experimental one.

The study is based on the analysis of the globally coupled chaotic systems [7]. It shows that tiny phase differences among the elements are amplified through chaotic dynamics, which then leads to dynamical clustering of the elements. The temporal pattern of the clustering is robust against external noise or is deterministic even if differences in the initial state are given stochastically. Our model takes into account of the feature of the globally coupled chaotic system, while the change in the degree of freedom is brought about by the cell division in the course of cell differentiation, which is also noted as a novel dynamical systems problem termed open chaos [8, 9].

Although we hope to reinterpret the cell biology [12] from our inter-intra dynamics viewpoint, it may be interesting, for the moment, to point out two experimental results. Yomo and his colleagues reported that even under single external condition, the cells differentiate to some distinct physiological states [10]. In their experiment, it was shown that one cell type of E. coli could result to a population with several distinct cell types after successive cultivation in a well-stirred liquid culture to impose the same external condition on each of the cells. Thus, even under the same initial and external condition, the cells can autonomously differentiate.

Rubin [11], in a series of papers, has shown that a cell line from mouse epigenetically transforms to different types of foci in size under the same condition. In addition, the frequency of transformation and types of the transformed cells were shown to depend on the cell density and the history of the cell culture. This suggests that transformation or differentiation of cells is dynamically generated by inter-cellular interaction.

3. Isologous Divisersification

From several simulations of the model starting from a single cell initial condition, Yomo and the author proposed the following "isologous diversification theory". It is a general mechanism of spontaneous differentiation of replicating biological units [1, 2, 3].

(1) Synchronous oscillations of identical units

Up to some number of cells, the biochemical oscillations of all cells are coherent, and they have almost same concentrations of chemicals. Accordingly, the cells divide almost simultaneously, and the number of cells is the power of two.

(2) Differentiation of the phases of oscillations of internal states.

When the number of units exceeds the threshold, they lose identical and coherent dynamics. Cells separate into several groups whose phases of oscillations are close. At this stage, only the phases of oscillations are different by cells, but the temporal averages of chemicals, measured over some periods of oscillations, are almost identical. The behavior here is nothing but the clustering of phases studied in coupled nonlinear oscillators. As has been discussed [8, 1], this temporal clustering corresponds to time sharing for resources: Cells can get chemical resource successively in order with the use of difference of phase of oscillations, because the ability to get it depends on the chemical activities of cells.

(3) Differentiation of the amplitudes of internal states.

After some divisions of cells, differences in chemicals start to be fixed by cells. The average chemical concentrations and their ratios differ by cells, even after taking the temporal average over periods. Thus the behavior of states is differentiated. The orbits of chemical dynamics lie in a different region by groups within the phase space.

It is also interesting to note that the frequency of oscillations is also differentiated. One group of cells oscillates and divides faster than the other group. Hence the differentiation of inherent time scales of cells emerges spontaneously through cell divisions.

(4) Transfer of the differentiated state to the offsprings by reproduction.

After fixed differentiation, chemical compositions of each group are inherited by their daughter cells. After some divisions, cell's chemical composition remains to be almost same, and thus the cells keep the "recursivity" by divisions.

It is important to note that the chemical characters are "inherited" just through the initial conditions of chemical concentrations after the division, although we have not explicitly imposed any external mechanisms specific to each cellular differentiation process. The determination of a cell has occurred at this stage, since daughters of one type of cells preserve the type. By reproduction, the initial condition of units is determined to give next generated units of the same type. Thus a kind of memory is formed, attained through the transfer of initial conditions on chemical concentrations by the cell division.

The cellular memory at this fourth stage is formed as the result of the selection of initial conditions for a cellular state (i.e., a partial system of the total dynamical system). As for the choice of initial conditions of the internal cellular system, this selection could be related with the basin for multiple attractors. However, in our model, each cellular state is not a stable state by the cell itself, but is stabilized through cellular interactions. The observed memory lies not solely in the internal states but also in the interactions among the units.

To see this inter-intra nature of the memory explicitly, one effective method is the transplantation experiment. Numerically, transplantation experiments are carried out by choosing determined cells (obtained at the normal diffusion process) and putting them into a variety of surrounding cells, that are not seen in the "normal" course of differentiation and development.

When a determined cell is transplanted to other cell society, the offsprings of the cell remain to be of the same type, unless the cell-type distribution of the society is strongly biased (i.e., the ensemble consisting of the same type of the cell as transplanted). Summing up several numerical experiments, we can conclude that the cell memory is preserved mainly in each cell, but cellular interactions are also important to sustain it. The achieved recursivity is understood as the choice of internal dynamics through cellular interactions. Thus the cellular interactions play the role not only of the trigger to differentiation, but also of the maintenance of diversity of cells. This is in agreement with the experimental results obtained in E. coli population [10]. Internal cellular memory is maintained as long as the diversity is sustained.

(5) Hierarchy of organized groups.

As the cell number increases, further differentiation proceeds. Each group of cells further differentiates into two or more subgroups. Some cells behave as a stem cell to support further differentiated cells. Often the switching of cell types to further subgroups are given by a stochastic automaton rule, where the stochasticity is supported by the chaotic intra-cellular dynamics[13]. It is also interesting to note that the rate of the differentiation to each subgroup depends on the cell-type distribution, so that the stability of the distribution of cell types is kept. Thus, the total system consists of units of diverse behaviors, forming a heterogeneous society.

4. Discussion

Here we have proposed isologous diversification theory for cell differentiation. So far there are two types of theories on the cell differentiation, one at a macroscopic level of Turing instability, and the other at a microscopic level of gene networks. Ours does not contradict with these previous theories, but is distinguished by providing new viewpoints as inter-intra dynamics. With it global information on cell distribution is embedded into intra-cellular dynamics. In contrast with the Turing instability mechanism, we explicitly take into account of internal degrees of freedom of a cell, while the mechanism leading to differentiation is temporal rather than spatial. As in the gene network theory, some characters are successively switched on. In contrast with the theory, however, this switching-type expression is not programmed explicitly but emerges as a dynamical mechanism.

In essence, the isologous diversification consists of the following triplets

(1) heterogeneity induced by clustering due to orbital instability in each internal dynamics

(2) stability as an ensemble level, provided by collective dynamics of coupled nonlinear elements

(3) recursivity as choice of initial conditions with digitalization of differentiated states

As to the first point, orbital instability in dynamical systems is essential. Tiny difference between two cells is amplified dynamically. As a corollary of the point 1), relevance of chemicals with low concentration is expected. In our simulation, chemicals with tiny amounts in cells are relevant to the trigger to differentiations. At differentiated cells, the difference is most remarkable for chemicals with low concentrations. The relevance is again a consequence of the amplification of tiny difference by nonlinear mechanism; difference of "rare" chemicals by cells can easily be amplified to lead to a macroscopic difference of cells. It should be noted that this relevance of chemicals with low concentrations is also supported in physiological facts.

The stability at a macroscopic level (2) has been discussed in dynamical systems theory, where dynamics of an ensemble of chaotic elements keeps some stability through the interaction [14, 15]. In our simulation, removal of some number of cells of a given type enhances the growth speed of the removed type, which restores the original ratio of types of cells. An extreme example is seen in the transplantation experiment, where elimination of all cells of one type leads to the transformation of the other type of cell.

With regards to the third point, the formation of digital memory is essential. Since the earlier stages, the phase of oscillations differs by cells. This phase difference is given by "analogue" means, and cellular states at any phase of oscillations can exist in principle. On the other hand, the differentiation based on the amplitude level is digital, in the sense that only discrete levels of amplitudes are allowed. Hence there are two levels of differences by cells, one for the change of phases of oscillations, and the other for the fixed differentiation. However, the differences by phases of oscillations are not rigid, since the phase is easily diffused by external disturbances: Perturbations brought about by division are enough to shift the phase and destroy the memory of the previous clustering. On the other hand, the "digital" difference by the amplitude of oscillations is more rigid, since it is not shifted continuously as in the case of phase. This emergence of digital information is the basis of the cellular memory at the next stage.

This "digital" distinction of chemical characters, that has emerged at the third stage, is relevant to preservation of them to daughter cells, since analogue differences of phases may easily be disturbed by the division process, and cannot be transmitted to daughters robustly.

It should be noted that the proposed theory captures the essence of cell differentiation, such as the loss of totipotency, origin of stem cells, differences in growth rates, and the importance of small or minimal amount of chemicals that trigger differentiation.

Furthermore, when a suitable condition is lost a "tumor"-like cell is observed[3], which is characterized by (i) extraordinary differentiation (ii) the loss of internal chemical diversity, with a simpler ongoing chemical pathway (iv) faster growth than normally differentiated cells, and (v) destruction of the ordered use of chemical resources. Indeed the formation of tumor-type cells is a consequence of isologous diversification theory. In the theory the differentiation process is not programmed explicitly as a rule but occurs through the interaction. Thus, when a suitable condition of the interaction is lost, for example, by the increase of density, "selfish" cells destroying the cooperative use of resources can be formed. It is also interesting to note that the tumor cell is set back to normal by supplying chemicals, to recover the diversity.

We believe that our "isologous diversification" can generally be applied to a variety of biological systems, because it is based on our study of coupled dynamical systems, which is expected to be universal in a class of interacting, reproducing, and oscillatory units. In the theory, the differentiation is triggered by the amplification of microscopic difference and thus is affected by a noise, but the globally behavior of an ensemble of the units is robust against external perturbations.

The origin of multicellular organism, for example, is directly related with our scenario. For its origin, some mechanism of differentiation of identical cells is necessary which leads to divisions of labor. According to our results, this feature of a multicellular organism spontaneously emerges as a consequence of strongly coupled reproducing units. Such differentiation is necessary for cells to increase their number within limited resources. Indeed the number of cells is saturated at a small level in our simulation, when the parameters of intra-cellular chemical oscillations are chosen so as not to allow for the clustering of oscillations.

acknowledgments

The present paper is based on the collaborate study with T. Yomo and C. Furusawa. I am grateful to them for stimulating discussions. The work is partially supported by Grant-in-Aids for Scientific Research from the Ministry of Education, Science, and Culture of Japan.

References

[1] K. Kaneko and T. Yomo, Physica 75 D (1994), 89

[2] K. Kaneko and T. Yomo, in *Advances in Artificial Life"*, Springer (1995) 329 (eds. E. Moran et al.)

[3] K. Kaneko and T. Yomo, Bull. Math. Biol. 59 (1997) 139

[4] B. Goodwin, "Temporal Organization in Cells", Academic Press, London (1963).

[5] B. Hess and A. Boiteux Ann. Rev. Biochem. 40 (1971) 237

[6] S.A. Kauffman, J. Theo. Biology. 22, (1969) 437

[7] K. Kaneko, Phys. Rev. Lett. 63 (1989) 219; Physica 41 D (1990) 137; Physica 54 D (1991) 5

[8] K. Kaneko, Physica D (1994) 55

[9] K. Kaneko, Artificial Life 1, (1994) 163

[10] E. Ko, T.Yomo, I. Urabe, Physica 75D (1994) 84

[11] A. Yao and H. Rubin,, Proc. Nat. Acad. Sci. 91 (1994) 7712; M. Chow, A. Yao, and H. Rubin, ibid, 91(1994) 599; H. Rubin, ibid, 91(1994) 1039; 91(1994) 6619

[12] e.g., B. Alberts, D.Bray, J. Lewis, M. Raff, K. Roberts, and J.D. Watson, *The Molecular Biology of the Cell*, 1983

[13] C. Furusawa and K. Kaneko, in preparation

[14] K. Kaneko Phys. Rev. Lett. 65 (1990) 1391; Physica 55D (1992) 368

[15] K. Kaneko and T. Ikegami,Physica 56 D (1992) 406

Pattern Recalling Property of Neural Network Modules Integrated by Chaos

Akira Sano

School of Information Science, Japan Advanced Institute of Science and Technology(JAIST),
Tatsunokuchi, Ishikawa 923–12, JAPAN

Abstract

The role of chaotic behavior in neural network model has been studied by several researchers. We bring forward an evidence that the chaotic neural network modules can be integrated without controller. In our two-body neural network model, the chaotic behavior plays the role of a mechanism that integrates network modules. Integration of modules was realized by time-shared pattern recalling. This result provides another viewpoint of chaos in neural network model.

Key words: Chaos in neuronal systems, integrating modules, time-shared pattern recalling

1 Introduction

A distinctive feature of information processing in a brain is the distributed and integrated process of good many functional modules, as cortical column, which were organized by sensory inputs. However, the existence of a control region to integrate modules has not been recognized. We consider the functional integration with modules can be realized without a controller in the cerebral neocortex, rather autonomously. In this article, we investigated the possibility of the functionally and autonomously integrating of modules on a neural network model.

It is known that the chaotic behavior plays some important roles in neural networks (and real neuronal systems) [1]–[4]. They are emphasized as regards on the following: [5] 1. a novelty filter, 2. explorative deterministic noise, 3. a fundamental form of neural activity that provides continuous, sequential access to memory patterns, and 4. a mechanism that underlies the formation of complex categories.

We investigate a chaotic neural network model that has multiple modules for the purpose of adding a role of chaos to the above opinions. We study our model on the assumption that chaotic behavior will perform to integrate the neural network modules without controller.

2 Flow-typed Two-Body Nozawa Model

Multi-body neural network model [6] is a simple framework formed by unidirectional (or bidirectional) inter-coupling the parts of conventional neural networks that several researchers have studied so far.

Two-body Nozawa model is given as a fundamental structure of the multi-body neural network model based on the Nozawa model [7] having chaotic behavior. The two-body Nozawa model is defined by adding external coupling terms to the Nozawa model as follows:

$$p_i(n+1) \quad = \quad F_{q_i(n)}\{p_i(n)\} \tag{1}$$

$$q_i(n) \quad = \quad \frac{1}{T}\left\{(1-\epsilon)\sum_{j\neq i}^{N} T_{ij}p_j(n) + \epsilon \sum_{k=1}^{M} T'_{ik}p'_k(n) + I_i\right\} \tag{2}$$

$$where \quad F_q(p) = rp + (1-r)\left[1 - \frac{1}{2}\left\{1 + \tanh\left(\frac{p-q}{2\beta}\right)\right\}\right] \tag{3}$$

In regard to Eq.(2) that includes external coupling terms added to a multi-body (two-body) Nozawa model, $p'_k(n)$ denotes an internal buffer of elements of external modules, T'_{ik} denotes a coupling weight from elements of external modules, and ϵ and M denote coupling ratio and the number of coupled elements, respectively.

A result of the *flow-typed* two-body Nozawa model is indicated in this article. The term, *flow-typed* means unidirectional inter-module couplings. The influence of inter-module inputs can be observed more purely in this flow-typed model.

The inter-module connections and its inputs act as perturbations on a module that is received inter-module inputs. When modules of a multi-body system do not form a multi stable system, and have a chaotic behavior, just extremely weak inputs (perturbations) from other modules will give strong influence. The behavior of the module after the perturbation from external inputs, is driven by just an internal rule of the modules.

3   Integrating Patterns on Chaotic Modules with External Inputs

The two modules of two-body Nozawa model are called module I and II, respectively. Now, two sets of three orthogonal patterns are prepared for each module, $\{C, F, 4\}$ and $\{C', F', 4'\}$. The prepared sets of orthogonal patterns are embedded into the each module using Hebb rule. The Hebb rule determines the coupling weight T_{ij} between the neuron elements within the modules. Similarly, the couplings between the modules T'_{ik} are also given by Hebb rule. Then, T'_{ik} give the matching of patterns between module I and II (e.g. $C \Leftrightarrow C'$).

Furthermore, the inter-module couplings are restricted unidirectionally from module II to I.

We have studied pattern recalling characteristics of the flow-typed two-body Nozawa model, in regard to change of two parameters for the inter-module coupling density in Eq.(2) ; $0 \leq M \leq 16$ and $0 \leq \epsilon \leq 1.0$. In this short article, we illustrate one of the recalling properties of the flow-typed two-body Nozawa model with external inputs. The external input is the pattern C on module I, and the pattern F' on module II (Fig. 2).

We have already known that the recalling on the Nozawa model, that is a single-body chaotic neural network model, wanders randomly from pattern to pattern embedded by the Hebb rule, when there is no external input [7].

We have observed pattern recalling ratio in regard to a M-ϵ parameter space. The recalling pattern of module I is completely in C without a inter-module flow input. Then, it can be seen that the pattern recalling ratio in module I is destroyed rapidly with a small increase of M. However, with small ϵ, the recalling of pattern C is preserved regardless of the value of M.

On the other hand, the recalling pattern F occurs as the inter-module input increases in wide region of the M-ϵ space, due to the projection from module II which recalls pattern F'.

Therefore, both embedded pattern C(with external input) and F(with following the intermodule coupling) have relatively large recalling ratio with small values of ϵ and middle values of M with module I. Figure 1 shows the recalling pattern sequence of module I in that parameter region; $(M = 7, \epsilon = 0.2)$.

The large recalling ratio of both the pattern C and F together is caused by a way of pattern recalling as time-shared (Fig. 1). However, on the other region of M-ϵ parameter space, one of two patterns are destroyed.

On the parameter region with recalling both C and F, our two-body system has the cooperative pattern recalling as time-shared, without controller.

We consider that it is an important role of chaos for information integration and temporal cording. The above results show an instance of that chaotic behavior in the two-body neural network is useful for the functionally integration as for information(pattern) processing.

A. Sano

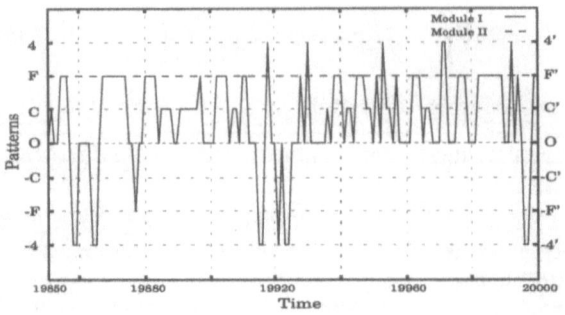

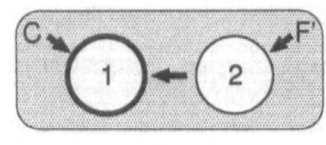

Fig.1: Recalling Sequence of a Two-Body Flow-typed Nozawa Model: $M = 7, \epsilon = 0.2$

Fig.2: Conceptual Diagram of Flow-typed Two-body Nozawa Model with External Inputs

4 Conclusions

The flow-typed two-body Nozawa model that is a fundamental structure of the chaotic multi-body model has been analyzed. Hence, we have shown a case where the chaotic behavior can be employed for integrating neural network modules without a controller. The pattern integration is realized as time-shared by the Hebbian synaptic connections.

In addition the Happel&Murre's four opinions[5], our results suggest a new role of chaotic behavior on neural network models to integrate the neuronal modules. Moreover, this idea with chaos might generally applicable for integrating functional subsystems.

References

[1] Y. Yao and W.J. Freeman. Model of biological pattern recognition with spatially chaotic dynamics, *Neural Networks*, Vol.3, pp.153–170, 1990.

[2] I. Tsuda. Dynamic link of memory – chaotic memory map in nonequilibrium neural networks, *Neural Networks*, Vol.5, pp.313–326, 1992.

[3] M. Adachi, K. Aihara, and M. Kotani. Nonlinear Associative Dynamics in A Chaotic Neural Network, In *Proc. of the 2nd International Conference on Fuzzy Logic & Neural Networks*, pp.947–950, 1992.

[4] S. Nara, P. Davis, and H. Totsuji. Memory search using complex dynamics in a recurrent neural network model, *Neural Networks*, Vol.6, pp.963, 1993.

[5] B.L.M. Happel and J.M.J. Murre. Evolving complex dynamics in modular interactive neural networks, 1996, (preprint).

[6] A. Sano. The role of intersubsystem coupling density and chaotic dynamics in multi-body neural network model, In *Proc. of the IIW95*, TR–95010 of *RWC Technical Report*, pp.41–53, RWCP, 1995.

[7] H. Nozawa. A Neural Network Model as a Globally Coupled Map and Applications Based on Chaos, *Chaos*, Vol.2, pp.377–386, 1992.

A complexity measure of the Internet

Osamu Yamakawa[1], Takashi Mori[3], Ryoku Nakamura[2], Kiyoshi Kudo[3], Yoichi Tamagawa[3],
Hidetoshi Suzuki[3]

1 Center for Information Science, Fukui Prefectural University, Matsuoka, Fukui 910-11, Japan
2 Department of Physics, Fukui Prefectural University, Matsuoka, Fukui 910-11, Japan
3 Department of Applied Physics, Fukui University. Fukui 910, Japan

Abstract

The diversity of entropy, as a measure of complexity, for data traffic on Internet has been studied. We also calculated the measure of artificial sequences , such as periodic, random, mixed, cellular automata generated and logistic map generated sequences, to compare with that of data traffic on the Internet. As a result, we find, the measure of data traffic is larger than those produced by other artificial sequences.

Keywords: Complexity, Diversity, Internet, Data traffic, Entropy

1. Introduction

The main aim of physics is explain nature surrounding us by making mathematical or algorithmic models. And then quantitative measures which characterize nature are needed to compare the models and nature. For studying "Complex Systems", measures of complexity, which characterize the systems, are also needed to relate models to real complex systems. So far, much effort has been made to investigate the measures of complexity.[1, 2] However, these measures have not been applied to real systems but artificial systems. We apply the diversity of entropy, as a measure of complexity, to the data traffic on Internet and some artificial systems to compare the complexity of them.

2. Diversity of Entropy

The temporal diversity of system behavior will be large for complex systems. Therefore, we measure the diversity of entropy for time sequences produced by real and artificial systems.

We get a series of real numbers as time sequence produced by systems, and divide that to several sub sequences length m for calculating a diversity of entropy with each sub sequence. The sub sequence $\{s_i, s_{i+1}, ..., s_{i+m-1}\}$ is converted to a binary sequence $\{b_i, b_{i+1}, ..., b_{i+m-1}\}$ by setting a threshold value which is median of s_i in the sub sequence. We call $p\{B\}$ the probability to observe a sub-sub sequence $B = \{b_k, b_{k+1}, ..., b_{k+n-1}\}$ of length n in a sub sequence $\{b_i, b_{i+1}, ..., b_{i+m-1}\}$ of length m $(n << m)$. The entropy of sub sequence length m is

$$H_m = -\sum_B p\{B\} \log p\{B\}$$

We observe several entropy of sub sequence in a whole sequence, thus, the diversity of entropy is defined as

$$\sigma_H = \sqrt{< H_m^2 > - < H_m >^2}$$

where $< A >$ denotes average of A.

3. Data traffic on Internet

On the Internet, all data are transferred within packets. Thus, we record the packet transferred time and the packet size by inserting a traffic monitor at the Ethernet segment, which is connected to the Internet through a router, in Fukui Prefectural University LAN(Fig.1). On this LAN, 300 of personal computers, 20 of workstations and 5 of file servers are connected. E-mail, Netnews, IRC, WWW, Ftp and Telnet are used on the network by 1500 of students and 200 of faculties. The data were recorded in a period over 24 hours. From this data, traffic density data(TD) is calculated by adding packet sizes in every second. Packet interval data(PI) is also generated to put interval time, between one packet and next packet, in order of time. TD shows self-similarity structure (Fig.2) which is pointed out by Leland et al.[3]. However, the slope of power spectrum for PI is nearly flat (Fig.3). This shows the sequence of packet interval is close to random.

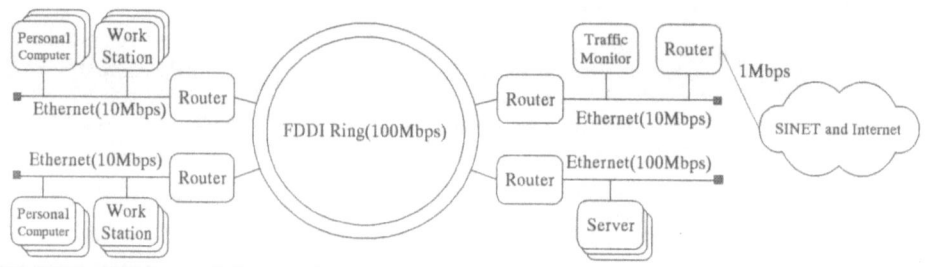

Fig.1.   Fukui Prefectural University LAN on which traffic data are taken.

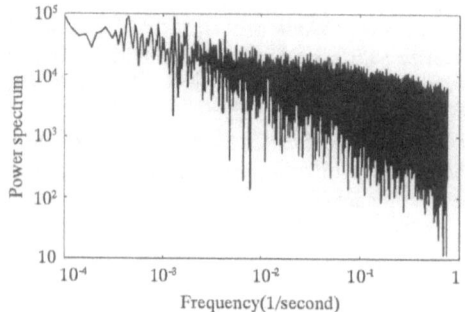
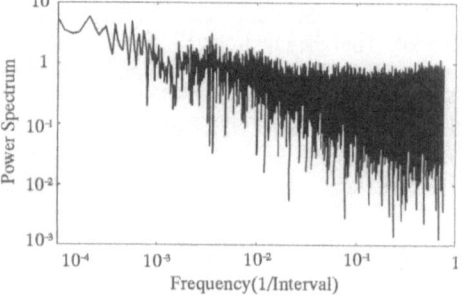

Fig.2. Power spectrum of traffic density.   Fig.3. Power spectrum of packet interval.
Fitted slope is -0.497.   Fitted slope is -0.107.

4. Other artificial sequences

Cellular automata has been investigated to explore and explain the complexity[4]. The diversity of entropy, as a complexity measure, is applied to the systems for comparing that of the Internet. The model of cellular automaton, we use, is one-dimensional elementary type. The rule 22(CA22) and the rule 54(CA54), which are classified the class III and the class IV[5, 6], are used to measure the diversity of entropy.

Logistic map has been used to study chaos and complex systems[7]. Thus, we also apply the diversity of entropy to the sequences which produced by logistic maps. This measurement are made at the control parameter as 3.7, 3.8284 and 4.0 which are called simple chaos(LM37), edge of chaos(LM38) and fully developed chaos(LM40), respectively.

This measure of complexity is also tested by a random sequence(RS), a periodic sequence(PS) and mixed sequences(MS), because we expect this measure becomes small on both random and periodic sequences.

5. Results and Discussions

The σ_H for the traffic density, packet interval of the Internet and several artificial sequences are shown in Table 1. We use the parameters of $n = 10$, $m = 600$ and the length of a sequence is 86400. From this point of view, the order of complexity for these real and artificial sequences is TD>LM38>CA54>LM37>MS>PI>CA22>LM40>RS>PS. We can easily accept this order of complexity by our intuition. Thus, the diversity of entropy(σ_H) is a good measure of complexity according to our feelings.

Table 1: Diversity of entropy(σ_H) for traffic density, packet interval and several artificial sequences.

	TD	PI	PS	RS	MS	CA22	CA54	LM37	LM38	LM40
σ_H	0.97	0.19	0.015	0.047	0.23	0.14	0.44	0.25	0.51	0.094

TD and LM38 have big complexity values, however, the structure of sequences are different each other. TD has a nature of self-similarity, though, LM38 doesn't. The sequence of LM38 has random parts and periodic parts, and these two parts come up in turn. CA54 has also a big complexity value, and the structure of sequence is same as LM38. This means that production mechanisms of complexity are different between real sequences and artificial sequences which are called complex. The diversity of entropy is good for distinguishing these two structure, however, we don't know what kind of real or artificial sequences make this complexity maximum. Thus, the nature of this measure should be study in next step.

To see the traffic density on local LAN is to see the local activity of the Internet. For studying the global activity or crowdeness of trunk line, the "ping" command, which measure a round trip time of packet between a host and target computer[8], should be used.

References

[1] P.Grassberger, Toward a Quantitative Theory of Self-Generated Complexity, *International Journal of Theoretical Physics*, Vol.25, No.9, 1986, pp.907-938.
[2] R.Wackerbauer, A.Witt, H.Atmanspacher, J.Kurths, H.Scheingrber, A Comparative Classification of Complexity Measures, *Chaos, Solitons & Fractals*, Vol.4, No.1, 1994, pp.133-173.
[3] W.E.Leland, M.S.Taqqu, W.Willinger and D.V.Wilson, On the Self-Similar Nature of Ethernet Traffic(Extended Version), *IEEE/ACM Transactions on Networking*, Vol.2, No.1, 1994, pp.1-14.
[4] S.Wolfram, Statistical mechanics of celluar automata, *Reviews of Modern Physics*, Vol.55, No.3, 1983, pp.601-644.
[5] S.Wolfram, University and complexity in cellular automata, *Physica D*, Vol.10, 1984, pp.1-35.
[6] W.Li and N.Packard, The structure of the elementary cellular automata rule space, *Complex System*, Vol.4, No.3, 1990, pp.281-297.
[7] M.J.Feigenbaum, Quanitative universality for a class of nonlinear transformations, *Jounal of statistical physics*, Vol.19, pp.25-52.
[8] I.Csabai, 1/f noise in computer network traffic, *Journal of Physics A: Mathematical & General*, Vol.27, 1994, pp.L417-L421.

Complexity and Diversity for the Rule Changeable Elementary Cellular Automata

Takashi Mori[1], Kiyoshi Kudo[1], Yoichi Tamagawa[1], Ryouku Nakamura[2], Osamu Yamakawa[3], Hidetoshi Suzuki[1], Tadayuki Uesugi[1]

1 Department of Applied Physics, Fukui University, Fukui 910, Japan
2 Department of Physics, Fukui Prefectural University, Matsuoka, Fukui 910-11, Japan
3 Center for Information Science, Fukui Prefectural University, Matsuoka, Fukui 910-11, Japan

Abstract

The rule changeable elementary cellular automata are proposed analogous to the ecological system. In this model we find the behavioral diversity for a suitable threshold value x on chaotic source rule. The behavioral diversity is characterized by block entropy and joint entropy. Our behavioral diversity plays just as "*edge of chaos*" for spatial and temporal evolution.

Keywords: Rule change, Cellular automaton, Edge of chaos, Complexity, Diversity

1. Introduction

We investigate the elementary cellular automata [1-2] on the ecological viewpoint. In ecosystem there exist various interactions among many species. Each species gets professional role to coexist one another. Thus behavioral diversity is appeared. Such diversity is one feature in complex systems. From this viewpoint we propose the rule changeable elementary cellular automata. We regard a "*rule*" in cellular automata as a "*species*" in ecosystem. Various interactions are brought about by the rule change. In this model we investigate the dynamics of spatial and temporal evolution and measure the behavioral diversity.

2. Model

We propose the rule changeable elementary cellular automata analogous to the ecological system. In elementary cellular automata, each cell has "*source rule*" which is the relationship between inputs and their inherent outputs. In our model, the rule change is brought about by introducing the variable of "*input frequency*". When the variable of input frequency vanishes, the input is captured to the similar input whose variable of frequency is high. We call the former "*captured input*" and the latter "*master input*". Cosequently, the output of the "*captured input*" changes to the output of the "*master input*". Further, when the variable of "*master input*" frequency vanishes, he is captured to the other input together with his "*captured inputs*". The "*captured input*" can recover his inherent output if his variable of frequency goes up to a certain threshold value x, which is the most characteristic parameter in our model. In this way, the relationship between inputs and their outputs changes succesively. Thus the rule change is brought about.

3. Results

Among many source rules, we apply our model to the rule 22 which is known to produce the most interesting feature [3]. We quantify each cell's behavior using block entropy [3] to observe the behavioral diversity. The diversity is measured by the disparsion of block entropy. As shown in Fig.1, average block entropy for all cells is high and the disparsion is low for low x (=0.0432). We plotted in Fig.2 block entropy v.s. joint entropy to observe the more detailed cell's behavior. In this case, all cells concentrate in the upper right region. This means that all cell's behavior is disordered. For medium x (=0.131), average block entropy is medium and the disparsion is high. In Fig.2, it is understood that the distribution of cells is very large. We conjecture that the behavioral diversity is appeared in this region. Besides, we find that as shown in Fig.3, its behavior shows "*edge of chaos*" which is recognized as the most complex behavior for the cellular automata [4]. For high x (=0.875), average block entropy and the disparsion are small. In Fig.2, all cells concentrate in the point where both entropy are zero. This means that all cell's behavior is ordered.

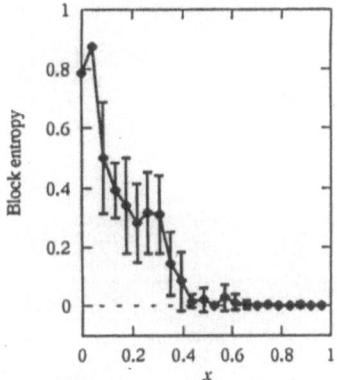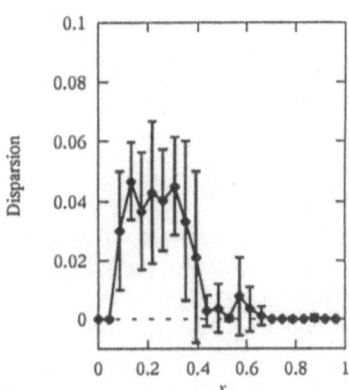

Fig. 1. The dependence of block entropy and the disparsion for the threshold value x on rule 22.

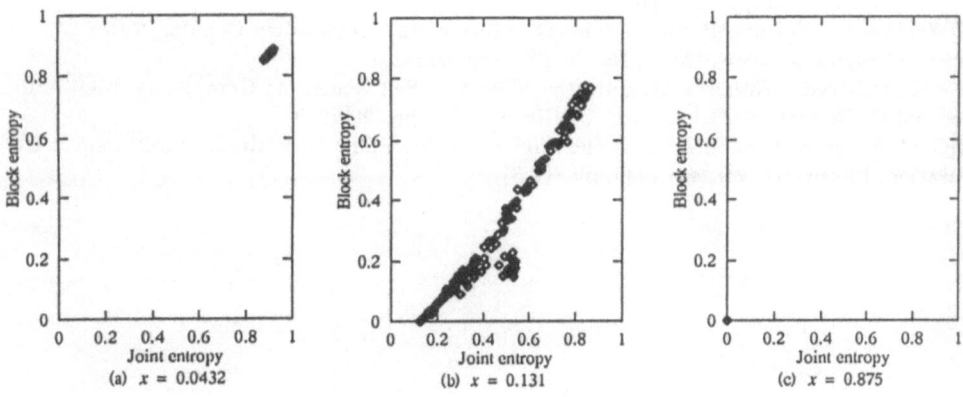

Fig. 2. The distribution of the detailed cell's behavior on rule 22.

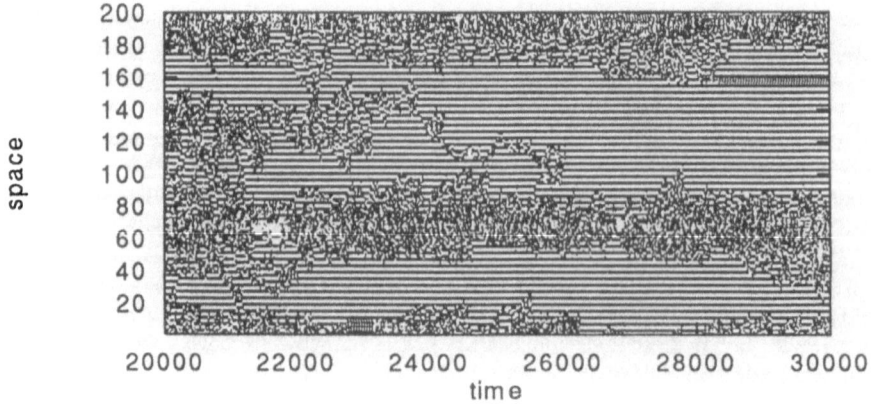

Fig. 3. The behavior on rule 22 at threshold value $x = 0.131$ from 20000 step to 30000 step per 10 step.

4. Discussion

In the present paper we have discussed the dynamics of the rule changeable elementary cellular automata for spatial and temporal evolution. In a periodic or fixed point source rules, the observed behavior is almost fixed point or periodic independent of threshold value x. In a chaotic source rules, the various behavior is observed. In the case of rule 22, we confirmed that the chaotic state turns to the ordered state with increasing x and the behavioral diversity, which shows just as "*edge of chaos*", is observed for medium x. By using block entropy and joint entropy, the behavioral diversity can be seen clearly.

References

[1]Stephen Wolfram. Statistical Mechanics of Cellular Automata, Reviews of Modern Physics, vol. 55, No. 3, 1983, pp. 601-644.
[2]Wentian Li, Norman Packard. The Structure of the Elementary Cellular Automata Rule space, Complex Systems, vol. 4, No. 3, 1990, pp. 281-297.
[3]P. Grassberger. Toward a Quantitative Theory of Self-Generated Complexity, International Journal of Theoretical Physics, vol. 25, No. 9, 1986, pp. 907-938.
[4]C. G. Langton. Computation at the Edge of Chaos: Phase Transitions and Emergent Computation, Physica D, vol. 42, 1990, pp. 12-37.

Generation of Relations between Individuals Using a Stochastic Automaton and the Theory of Change of Attitudes

Tatsuya Nomura

ATR Human Information Processing Research Laboratories, Soraku-gun, Kyoto 619-02 JAPAN

Abstract

We regard internal states of an automaton as emotional states of an individual, its outputs as its actions towards another individual, and a parameter affecting its state transition as a personal trait. We therefore propose a representation of an individual with emotional states and a certain personality, by a stochastic automaton and a model of a group in which each individual begins conversations with another individual. The relations between individuals are generated from results of the conversations based on the congruity theory in social psychology.

Key words: Stochastic Automatons, Emotion, Personality, Change of Attitudes

1 Introduction

Since the 1960s, many studies on computational models for emotion have been proposed for the purpose of representing the emotional process in one individual[1][2]. In these studies, however, neither groups of individuals with emotion nor relations between individuals or constraints (from relations in individual dynamics) have been dealt with.

In Artificial Life, on the other hand, groups that consist of individuals with simple dynamics, relations generated through communications between the individuals, and constraints from relations in individual dynamics play major roles. However, emotional states and personalities of individuals have not been dealt with, although some studies on emotion have employed mutual actions between entities[3][4].

In this paper, we propose a simple representation of an individual dynamics by using a stochastic automaton and a model of a social group consisting of individuals represented by automatons. In particular, we focus on relations between individuals and on generating these relations based on an analogy from the theory on attitude changes in social psychology.

2 Representation of Individuals based on a Stochastic Automaton

Let $E = \{E_1, E_2, \ldots, E_m\}$ be a set of symbols that represent emotional states and $A = \{A_1, A_2, \ldots, A_m\}$ be a set of symbols that represent actions. We define our stochastic automaton $AT(r, f)$ whose internal state space is E and whose input and output space is A as follows:

$$AT(r, f) = \{E, A, AC(r, \cdot, \cdot), EM(f, \cdot, \cdot), EF, \pi(f, r, \cdot)\}, \ r \in [0, 1]^D, \ f \in [0, 1], \quad (1)$$

$$AC : [0, 1]^D \times E \times A \to [0, 1], \quad EM : [0, 1] \times A \times A \times E \to [0, 1],$$

$AC(r, E_i, A_j)$: Probability that the output is A_j when the internal state is E_i,

$EM(f, A_j, A_k, E_l)$: Probability that the next internal state is E_l

when the input is A_j and the output is A_k,

$EF(\in E)$: Halting state,

$\pi(f, r, E_i)$: Probability that the initial internal state is E_i

$$\sum_{j=1}^{n} AC(r, E_i, A_j) = 1, \quad \sum_{l=1}^{m} EM(f, A_j, A_k, E_l) = 1, \quad \sum_{i=1}^{m} \pi(F, r, E_i) = 1$$

$$(i = 1, \ldots, m, j, k = 1, \ldots, n)$$

Here, r represents the personality parameter and f represents the attitude towards an object for output. Let $em(t) \in E$, $ac(t) \in A$, and $in(t) \in A$ be the internal state, the output, and the input at a discrete time t respectively. Then, the relation between $em(t)$, $ac(t)$ and $in(t)$ is given as follows:

$$Prob(em(0) = E_i) = \pi(f, r, E_i), \ Prob(em(t+1) = E_i) = EM(f, in(t), ac(t), E_i) \quad (2)$$
$$Prob(ac(t) = A_j) = AC(r, em(t), A_j) \quad (i = 1, \ldots, m, \ j = 1, \ldots, n, \ t = 0, 1, \ldots)$$

We represent a conversation between an individual with personality r and another individual with attitude f by regarding $AT(r, f)$ as the conversational dynamics of the individuals. $em(t)$, $ac(t)$, and $in(t)$ are regarded as the emotional state of one individual, the action of one individual, and the action of the other individual at a time t respectively. Furthermore, we make each individual have a frustration affecting the attitude and changing the probabilities during the conversation.

3 Generation of Relations between Individuals based on the Theory of Change of Attitudes

The congruity theory predicts attitude changes in communications based on cognitive consistency. Cognitive consistency is based on the idea that humans have a basic requirement: they want to maintain consistency in their beliefs, attitudes, and actions for objects; if inconsistencies exist, they cause displeasure. Humans are motivated to reduce displeasure.

In the congruity theory or some theories on cognitive consistency, a person (P), a concept toward which P has an attitude (C), and a message source that refers to C (S) are dealt with. In these theories, the balance in the attitude of P for C (PC), that of P for S (PS), and the message of S for C (SC) are major problems.

The balance theory defines that if the sign of the product of PC, PS, and SC is positive, the situation is balanced; if it is negative, the situation is inbalanced.

In the congruity theory, the changes in values of PC and PS is determined to balance the situation based on the following equations (from [5]):

$$\text{The Change in } PS \ = \ W(SC, PC, PS) = \begin{cases} \frac{|PC|}{|PC|+|PS|}(PC - PS) & (SC > 0) \\ \frac{|PC|}{|PC|+|PS|}(-PC - PS) & (SC < 0) \end{cases} \quad (3)$$

$$\text{The Change in } PC \ = \ W(SC, PS, PC)$$

We regard the attitude of Ind_p toward Ind_q as the message from Ind_p to Ind_q and the attitude of Ind_q toward Ind_p as the message from Ind_q to Ind_p, and define the modification of the attitudes of $Ind_s (s \neq p, q)$ based on (3).

4 Simulations for a Concrete Model

We executed simulations for the following concrete model.

$$E \ = \ \{"anger", "hatred", "fear", "pleasure", "sadness", "neutrality"\}$$
$$A \ = \ \{"rebound", "cooperation", "disregard"\}$$

We set the above probabilities so that the internal states are easy to transit to "*pleasure*" when the attitudes toward the others are positive, and the actions "*cooperation*" are easy to be selected when the personalities are low. Moreover, we set the frustrations so that they frequently change in the low personalities. We let the longest conversation time between a pair of the individuals T be 5 or 15 and the conversations for all pairs of individuals be one round. The following tables show the average through 10^6 rounds in 10^2 trials for each case.

Table 1: The State Distribution of the Group Consisting of Three Individuals with the same personality

State	$T = 5$, Low r	$T = 15$, Low r	$T = 5$, High r	$T = 15$, High r
Not Separated	9.8	91.3	0.0	0.0
Separated	89.7	8.7	100.0	100.0

Table 2: The State Distribution of the Group Consisting of Individuals with Different Personalities (3 with Low r and 1 with High r, $T = 5$)

Not Separated	1(Low r) : 3	1(High r) : 3	2 : 2
0.0	0.3	97.3	2.4

As shown in Table 1, the group was separated when the personality was high. Moreover, although the low personality makes the individual produce positive attitudes, the group was separated in the case of the low personality and short conversation time because the frustration frequently changed and made the attitudes unstable. As shown in Table 2, however, the three individuals with the low personality conspired with each other by the effect of the individual with the high personality in spite of the short conversation time.

5 Conclusion

We proposed a model representing a group that consists of individuals with emotional states by using a stochastic automaton and an analogy of social psychology.

This model imitates only a part of the relationship between actions and emotional states in humans, and does it quite wildly. Furthermore, there is no guarantee that the properties in the model reflect some social phenomena in the real world. In order to reflect real social phenomena, we need to consider many levels of emergence: the relationship between person perception in individuals and generation of structures of society, the relationship between desires in individuals and survival strategies of individuals under structures of society.

References

[1] M. Itoh, M. Umemoto, A. Yamadori, T. Ono, A. Tokosumi, and K. Ikeda. *Emotion*, volume 6 of *Cognitive Science*. Iwanami Shoten Publishing, 1994.

[2] A. Tokosumi, N. H. Frijda, D. Moffat, C. Elliott, M. Toda, and R. Pfeifer. Special section 'cognitive science of emotion'. *COGNITIVE STUDIES: Bulletin of the Japanese Cognitive Science Society*, Vol.1, No.2, 1994, pp.3–57.

[3] T. Ono and S. Sato. A computational model of emotion using the mechanism of the immune system. *COGNITIVE STUDIES: Bulletin of the Japanese Cognitive Science Society*, Vol.2, No.3, 1995, pp.48–65.

[4] M. Toda and Y. Takada. *Emotion: The Innate Adaptive Software System That Drives Human Beings*, volume 24 of *Cognitive Science Series*. University of Tokyo Press, 1992.

[5] Y. Matsuyama. *An Overview of Contemporary Social Psychology*. Kitaoji Shobo Publishing, 1982.

Dynamic Learning and Retrieving Scheme Based on Chaotic Neuron Model

Haruhiko Nishimura, Naofumi Katada, and Yoshihito Fujita

Dept. of Information Sci., Hyogo Univ. of Education, Yashiro-cho, Hyogo 673-14, JAPAN

Abstract

A stimulus-response scheme is introduced in chaotic neural networks with synaptic plasticities and the processes of dynamic learning under external stimuli are investigated. Owing to the refractoriness and the time-hysteresis, memory fixing abilities of stimuli become much higher than those by the Hopfield neural network with / without stochastic activities and also have sensitive dependences on the strength of stimulation. These characteristics turn out to be supported with the chaotic activity by examining the relation between the refractoriness and the Lyapunov exponent during the engraving of stimuli on the network. The above results indicate a possibility of realizing the real-time learning mechanism against the external time series inputs, which is difficult for the static Hopfield model of associative memory. Furthermore, modeling of dynamic retrieval mechanism is attempted for diverse switching phenomena on the chaotic neural network with grown (fixed) synapses.

Key words: Neural network, chaos, stimulus, learning, retrieving

1. Introduction

Diverse types of chaos have been confirmed at several hierarchical levels in the real neural systems from single cells to cortical networks (e.g. ionic channels, spike trains from cells, EEG). This suggests that artificial neural networks based on oversimplified neuron models should be re-examined and re-developed. Chaos may play an essential role in the functioning of associative memory, beyond the frame of the Hopfield neural network with only equilibrium point attractors. To make this point clear, following the model of chaotic neural networks by Aihara et al. [1], we have studied the dynamic learning and retrieving features under external stimuli [2][3]. In this paper we briefly report a part of the research.

2. Models and Methods

The chaotic neural network (CNN) composed of N chaotic neurons is described as [1][4]

$$X_i(t+1) = f\left(\sum_{j=1}^{N} w_{ij} \sum_{d=0}^{t} k_f^d X_j(t-d) - \alpha \sum_{d=0}^{t} k_r^d X_i(t-d) - \theta_i\right) , \tag{1}$$

where X_i : output of neuron $i (-1 \leq X_i \leq 1)$, w_{ij} : synaptic weight from neuron j to neuron i, θ_i : threshold of neuron i, $k_f(k_r)$: decay factor for the feedback(refractoriness) $(0 \leq k_f, k_r < 1)$, α : refractory scaling parameter, f : output function defined by $f(y) = tanh(y/2\varepsilon)$ with the steepness parameter ε. By taking internal states to be $\eta_i(t+1) = \sum_{j=1}^{N} w_{ij} \sum_{d=0}^{t} k_f^d X_j(t-d)$

and $\zeta_i(t+1) = -\alpha \sum_{d=0}^{t} k_r^d X_i(t-d) - \theta_i$, Eq.(1) can be reduced to

$$\eta_i(t+1) = k_f \eta_i(t) + \sum_{j=1}^{N} w_{ij} X_j(t) \ , \tag{2}$$

$$\zeta_i(t+1) = k_r \zeta_i(t) - \alpha X_i(t) + a \ , \tag{3}$$

$$X_i(t+1) = f(\eta_i(t+1) + \zeta_i(t+1)) \ , \tag{4}$$

where a is temporally constant $a \equiv -\theta_i(1-k_r)$. The network corresponds to the conventional discrete-time Hopfield network when $\alpha = k_f = k_r = 0$ (Hopfield network point (HNP)). The deterministic dynamics of HNP can be replaced according to a stochastic rule

$$\begin{cases} Prob\{X_i = f(h_i)\} = g_H(|h_i|) \\ Prob\{X_i = -f(h_i)\} = 1 - g_H(|h_i|) \ , \end{cases} \tag{5}$$

where $h_i = \sum_{j=1}^{N} w_{ij} X_j - \theta_i$, $g_H(h) = 1/(1 + e^{-h/H})$ and H is a positive temperature constant. This network is known as stochastic neural network (SNN).

Under external stimuli(Se), Eq.(4) is influenced as

$$X_i(t+1) = f(\eta_i(t+1) + \zeta_i(t+1) + \sigma_i), \tag{6}$$

where $\{\sigma_i\}$ is the effective term by Se, and simultaneously synaptic weights change plastically according to Hebb-like rule

$$w_{ij}(t+1) = w_{ij}(t) + \beta X_i(t) X_j(t) \tag{7}$$

with the plasticity parameter β for the duration of external stimuli(Ts). The growth of w_{ij} is limited by the upper bound K_i of the norm $\|w_i\| = (\sum_j^N w_{ij}^2)^{1/2}$. The flow diagram of dynamic learning (memory fixing) processes is shown in Fig.1. We can evaluate changes of the Lyapunov exponents to the network at each interval $T_I - Ts$.

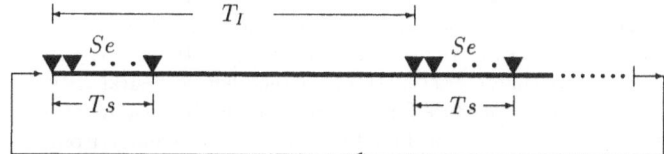

Fig.1. Flow diagram of stimulus-response scheme

3. Simulations and Results

To carry out computational experiments, we use the $12 \times 13(N = 156)$ non-orthogonal p alphabet patterns $\{\xi_i^\mu\}(\mu = 1, \cdots, p, ; i = 1, \cdots, N)$ as a set of external stimuli: $\sigma_i = s\xi_i^\mu$. s is the strength factor of stimulation. For time series evolutions of CNN, HNP and SNN under the scheme in Fig.1, we have analyzed the distribution of the final stability coefficients $\gamma_i^\mu \equiv \xi_i^\mu \sum_{j=1}^{N} w_{ij} \xi_j^\mu / \|w_i\|$ for patterns of stimuli (pattern's basin structure [5]), the final energy map $E = -(1/2) \sum_{ij} w_{ij} \xi_i^\mu \xi_j^\mu$ and so on. Figure 2 is a typical example of our results when $Ts = 10$, $T_I = 100$, $\beta = 0.1/N$, $s = 2.70$ and 8 patterns [R, Z, Q, Y, X, A, T, H] stimulate the network up to $t = 8000$. CNN is much superior to SNN in both holding efficiencies $|m^\mu|$ and depths of memory engraving $|E|$ for 8 pattern states. Here, $m^\mu = (1/N) \sum_{i=1}^{N} X_i \xi_i^\mu$.

We have also investigated dynamic retrieving features of CNN with grown (fixed) synaptic weights and observed transitions among the chaotic trembling modes (CTMs) which occur spontaneously or induced by external stimuli. CTMs are well-retrieval states, but with positive maximum Lyapunov exponent, i.e. $m^\mu \simeq 1$ and $\lambda_1 > 0$. Figure 3 shows the sensitive reactions of a CTM ($0.99 < m^F < 1.0$ and $\lambda_1 = 0.46$) to the external stimuli ($\{\sigma_i\} = 1.0, Ts = 1$). Different retrieving processes (to F, and to L) are caused by the same stimuli only with the difference of stimulation time.

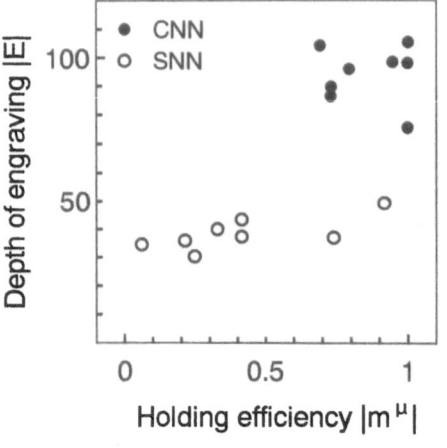

Fig.2. Comparison of fixing abilities between chaotic and stochastic neural networks.

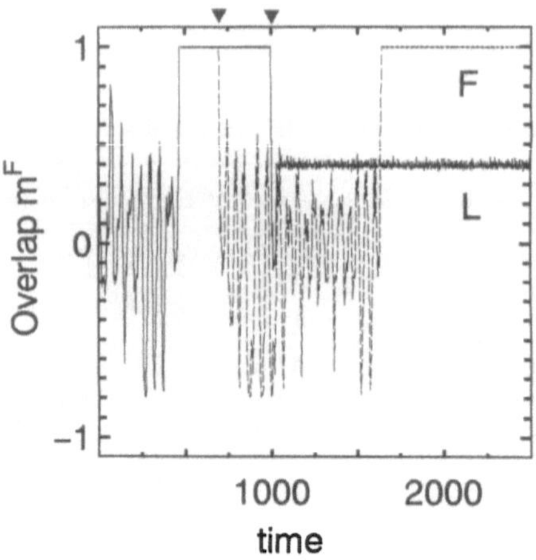

Fig.3. Time-sensitivity of a CTM reaction to external stimuli at different time

4. Concluding Remarks

We found that under the stimulus-response scheme in chaotic neural network with synaptic plasticities, the refractoriness much increases the efficiency of memory fixing. By examining the relation between the hysteretic refractoriness and the Lyapunov exponent, the chaotic activity seems to play an essential role to engrave the stimuli on the network. This is difficult for the stochastic activity to attain to. Diverse retrieval behaviors (chaotic itinerancy, chaotic trembling, chaotic burst) are observed as characteristic states on the chaotic neural network with grown(fixed) weights. By finding out a procedure for controlling high dimensional chaotic systems [6] from time series data, dynamic association may be understood to emerge from the above processes, not based on the elaborately programed rule.

5. References

[1] K.Aihara, T.Takabe, M.Toyoda, Chaotic Neural Networks, *Physics Letters*, A144, 1990, pp.333-340.
[2] H.Nishimura, Y.Fujita, S.Fujita, A Mechanism for Dynamic Association Based on Chaotic Neural Networks, *Proc. of 1994 Annual Conf. of JNNS*, 1994, pp.104-105 (in Japanese).
[3] H.Nishimura, N.Katada, Dynamic Learning Processes in Chaotic Neural Networks, *Proc. of 1995 Annual Conf. of JNNS*, 1995, pp.117-118 (in Japanese).
[4] M.Adachi, K.Aihara, An analysis of associatve transient dynamics in a chaotic neural network, in *Towards the Harnessing of Chaos*, Elsevier, 1994, pp.335-338.
[5] B.Müller, J.Reinhardt, *Neural Networks: An Introduction*, Springer-Verlag, 1990.
[6] K.Kaneko, Clustering, coding, switching, hierarchical ordering, and control in a network of chaotic elements, *Physica* D41, 1990, pp.137-172.

Genetic evolution of the subsumption architecture which controls an autonomous mobile robot

Ryoichi Odagiri [1], Taku Naito [1], Yutaka Matsunaga [2], Tatsuya Asai [1], and Kazuyuki Murase [1]

1 Dept., of Info. Sci., Faculty of Eng., Fukui Univ., Fukui-shi, Fukui 910 JAPAN
2 Faculty of Edu., Aichi Univ. of Edu., Kariya-shi, Aichi 444 JAPAN

Abstract

We propose a method to evolve the Subsumption Architecture to control autonomous robots. Each layers of the subsumption architecture was subdivided into sublayers. Each sublayer was composed of functional modules, which were determined by evolution in environment.

This algorithm was implemented in a real miniature mobile robot, Khepera, and several experiments were performed. Khepera successfully learned to navigate and avoid obstacles in test fields.

Keywords: Genetic evolution, autonomous behavior, mobile robot, subsumption architecture

1 Introduction

Recently, new approaches to built autonomous robots that can carry out useful work in unstructured environments have been developed [2, 6]. Among them, the subsumption architecture [1] was most frequently applied to control real robots, and seems most successful. In the subsumption architecture, the control system consists of layers which are built to let the robot operate at increasing levels of competence. Layers are made up of asynchronous modules, each of which is a computational machine to evoke action in response to sensory signals. Higher-level layers can subsume the roles of lower levels by suppressing their outputs. Lower levels continue to function even when higher levels are added. Thus, the designer of the controller could build up the layers in a step-by-step manner to achieve higher intelligence.

In the subsumption architecture, however, the structure of layers and the function of each layer have to be determined by the designer. At the beginning, he has to deal with a robot performing limited behaviors in environment, and gradually add up necessary layers in a trial-and-error manner. Genetic evolution of the subsumption architecture in environment has been proposed, and the computer simulation of a wall-following for a mobile robot was reported successful [4, 5] Since it was to determine the whole structure by the genetic algorithm regardless of prior knowledge on the control scheme, it took numerous generations to converge even for simple tasks.

We therefore utilized the original idea of the subsumption architecture; subdividing into layers by knowledge *in priori*, and determining each layer from lower to higher one by one. In order to further enhance the convergence, each layer was divided into sublayers with prior knowledge. The elements of a sublayer were determinined by evolution in environmnet.

This method was applied to a real mobile robot, Khepera. It successfully evolved to navigate and avoid obstacles. The evolution was faster than the previous method.

2 The Subsumption Architecture

The Subsumption Architecture has been compared with the conventional controller as shown in Fig. 1. In the latter, the problem is decomposed into a series of functional layers, such as sensing, modeling, planning, task execution, and motor control. A central system executes each layer, and passes the results onto the next layer in synchrony and a closely-coupled manner. In the former, the problem is decomposed in a set of layers which are asynchronous behaviors in themselves. The overall control is achieved by the collective effect of such local and primitive behaviors. Usually, higher layers subsume the lower layers, for example, a forward moving behavior is subsumed by a turning behavior to avoid an obstacle.

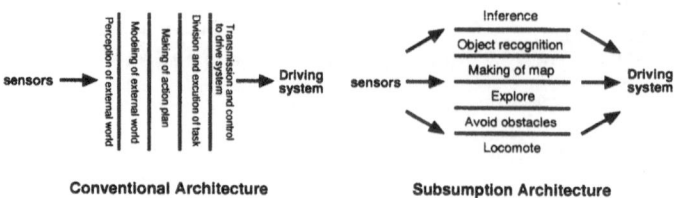

Conventional Architecture   **Subsumption Architecture**

Figure 1:

The subsumption architecture for navigation and avoiding obstacles could consist of two layers, the locomote layer and the avoid-obstacle layer. Thus, if the upper layer is activated, the lower layer is suppressed and the behavior of the upper layer is executed. Since the lower layer is to go forward, and if the upper layer is not activated, the robot continues to move forward.

3 Composing a layer with sublayers and a sublayers with functional modules

Our knowledge could be further utilized to divide each layer into sublayers. For example, the layer of "avoid obstacles" could be made up with three sublayers; follow the right wall, follow the left wall, and avoid head-on collisions, as shown in Fig. 2.

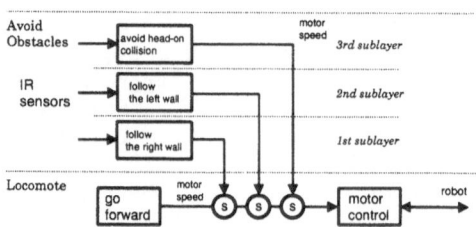

Figure 2: The subsumption architecture for navigation and avoiding obstacles

Each sublayer could be further divided into elementary behaviors that are evoked by a certain sensory information. We here call such elementary behaviors as "functional modules", that may be also called "sensor-motor reflexes". Functional modules could be described by the IF–THEN–UNTIL-type rules; IF a sensory pattern has occurred, THEN evoke an action UNTIL another sensory pattern occurs. Since they are basically simple reflexes, each functional module uses a limited sensory information and motor function.

4 Evolution of the architecture

Evolution could be performed from lower sublayers. Since each sublayer consists of a limited number of functional modules, each of which utilizes very limited sensory and motor functions, the number of freedom is low, and thus, the evolution is fast.

5   Experiments

The architecture as shown in Fig. 2 was implement into a mobile robot, Khepera. It had eight proximity sensors around the body and two motors that were individually controlled. Each sublayer was composed of two functional elements. Each of functional element utilized limited sensory information suitable for the elemental behavior. For example, the sublayer "Follow the right wall" used sensory input from sensors located at the right side of the body.

Result of the evolution were shown in Fig. 3. Each evolution converged within 20 generations, and desired behavior was acquired.

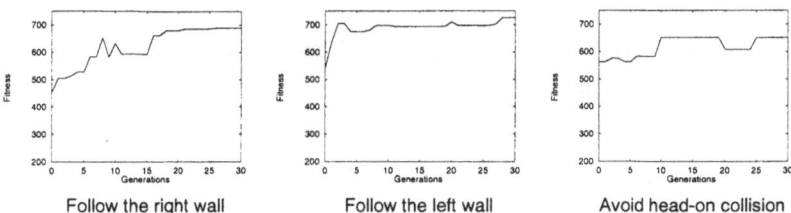

| Follow the right wall | Follow the left wall | Avoid head-on collision |

Figure 3: Genetic evolution

We also tried to evolve the architecture consisted of six functional modules at once. Each modules received input from all sensors. Even after 200 generations, the robot could not acquire the behavior.

6   Remarks

This method utilized our knowledge maximally to develop the control architecture, and only the unknown parts were determined by evolution in environment. Therefore, the evolution was faster and the final structure was reliable.

References

[1] Rodney A. Brooks. A robust layered control system for a mobile robot. *IEEE Journal of Robotics and Automation*, (RA-2):14–23, 1986.

[2] Rodney A. Brooks. New approaches to robotics. *Science*, (253):1227–1232, 1991.

[3] D. Floreano, F. Mondada, and P. Ienne. Mobile robot miniaturisation: A tool for investigation in control algorithms. *Proceedings of the third International Symposium on Experiment Robotics*, 1993.

[4] John R. Koza. Evolving emergent wall following robotic behavior sing the genetic programming paradigm. *Proceedings of the 1st European Conference on Artificial Life*, pages 110–119, 1991.

[5] John R. Koza. Evolution of subsumption using genetic programming. *Toward a Practice of Autonomous Systems: Proceedings of the first European Conference on Artificial Life*, 1992.

[6] S. Nolfi, D. Floreano, O. Miglino, and F. Mondada. How to evolve autonomous robots: Different approaches in evolutionary robotics. *Artificial Life IV Proceedings of the Fourth International Workshop on the Synthesis and Simulation of Living Systems*, 1994.

On the generation and dissipation of coherent structure in two-dimensional vortices

Yoshio ISHII

Dept. of Information Systems Science, Faculty of Engineering, SOKA Univ.,
1-236, Tangi-cho, Hachioji-shi, Tokyo 192, JAPAN.

Abstract

A vortex properties deduced has been used to validate a simple vortex model of turbulence, in which the vortices move like point vortices. The intention in this paper is to discuss the restricted subject of two-dimensional vortex motion. I deal with the dynamics of a system of point vortices in two-dimensional flow field and discuss a few points about generation and dissipation of coherent structure in these system.

Key words: Coherent Structure, long-range interaction, point vortex, turbulence

1. Introduction

In the last few years, the study of two-dimensional turbulence has received a rather large interest. Some of the results obtained by numerical simulation have been also observed in carefully designed laboratory experiments an two-dimensional turbulence. In particular, vorticity dynamics plays a major role in two-dimensional turbulence[1][2]. In generation and dissipation of coherent structure in two-dimensional turbulence, the emergence of coherent vortices has been observed with numerical simulations[3][4].

In this study, I describe generation and dissipation of coherent structure in two-dimensional turbulence using numerical simulation.

2. System of two-dimensional vortices

2.1 Point vortices

Point vortices offer a simplified description, valid when the vortices are concentrated and well separated. Vortices are then considered only in the form of singularities like point vortices and discontinuity sheets rather than as extended regions of vorticity. Such methods give insight into the behavior of vortex systems.

Figure 1. Deformation and transition in the system of vortices.

Figure 2. Merging and breakdown in the system of vortices.

A point vortex in the (x,y) plane, located at x_1, y_1, is described by

$$\frac{dx}{dt} = -k_1 \frac{y-y_1}{r^2} \qquad \frac{dy}{dt} = k_1 \frac{x-x_1}{r^2}$$

where $r^2 = (x-x_1)^2 + (y-y_1)^2$ and k is a strength of vortex.

The discussion is now extended to n point vortices. From the velocity components of the ith vortex

$$\frac{dx_i}{dt} = \sum_{k=1}^{n} k_k \frac{y_i-y_k}{r_{ik}^2} \qquad \frac{dy_i}{dt} = \sum_{k=1}^{n} k_k \frac{x_i-x_k}{r_{ik}^2}$$

with $r_{ik}^2 = (x_i-x_k)^2 + (y_i-y_k)^2$.

Except for a few special case the calculation of a system of vortices for n > 2 requires the aid of computers.

2.2 Taylor-Green Vortex

Taylor-Green Vortex is flow that develops the constant density incompressible Navier-Stokes equations with periodic boundary conditions. This is perhaps the simplest system in which to study the generation of excitation at small scales and the resulting turbulence.

3. Results and discussions

The results are shown in Figure 1 and 2. Figure 1 represents in the deformation of vortex and transition from vortex point to saddle point in the vortex system. Figure 2 consecutively represents in the merging (generation) and breakdown (dissipation) in the system of point vortices.

I expect that adding a time-dependent velocity field to the three-vortex system will in general produce chaotic dynamics[5]. A problem to be solved in this study, it is necessary to measure for the decision of various physical quantity. In particular, about entropy, power spectra, Lyapunov spectra . . . and so on. Moreover, it is important to investigate that chaotic itinerancy is or not appeared in this phenomena of long-range interaction between vortices.

References
[1] D. G. Dritschel, A Fast Contour Dynamics Method for Many-Vortex Calculations in Two-Dimensional Flows, Phys. Fluids, A5, No.1, 1993, pp.173-187.

[2] D. G. Dritschel, Vortex Properties of Two-Dimensional Turbulence, Phys. Fluids, A5, No.4, 1993, pp.984-997.

[3] P. Santangelo, R.Benzi, B. Legras, The Generation of Vortices in High-Resolution, Two-Dimensional Decaying Turbulence and the Influence of Initial Conditions on the Breaking of Self-Similarity, Phys. Fluids, A1, No.6, 1989, pp.1027-1035.

[4] R.Benzi, M. Colella, M. Briscolini, P. Santangelo, A Simple Point Vortex Model for Two-Dimensional Decaying Turbulence, Phys. Fluids, A4, No.5, 1992, pp.1036-1039.

[5] H. Aref, Integrable, Chaotic, and Turbulent Vortex Motion in Two-Dimensional Flows, Ann. Rev. Fluid Mech., Vol.15, 1983, pp. 345-389.

Dynamics of two coupled inverted pendulums

Hidetoshi Suzuki[1], Kiyoshi Kudo[1]. Yoichi Tamagawa[1], Ryouku Nakamura[2]
Osamu Yamakawa[3], Tadayuki Uesugi[1], Kazuhiro Kobayashi[1]

1 Department of Applid Phisics, Fukui University, Fukui 910, Japan
2 Department of Bioscience, Fukui Prefectural University, Matsuoka, Fukui 910-11, Japan
3 Center for Information Science, Fukui Prefectural University, Matsuoka, Fukui 910-11, Japan

Abstract

Dynamics of two coupled van der Pol type inverted pendulums are studied. A phase transition is observed between synchronized phase and asynchronized phase. Behavior of angular correlation near critical point is observed. Exact critical point and critical exponent are obtained by obeserving convergent time.

Keywords: Inverted pendulum, phase transiton, synchronization, angular correlation, critical exponent

1. Introduction

Interacting nonlinear oscillators have various behaviors. Drawing effect is one of their behaviors. It is very interesting phenomenon, since complex behaviors are given by small interaction. The purpose of this article is to observe and analyze drawing effect using coupling oscillators [1].
We investigate two coupled van der Pol type [2] inverted pendulums which give stable oscillations. This model present two phases, i.e., synchronized phase and asynchronized phase according to interaction parameter d. We get exact critical point and critical exponent.

2. Model

Inverted pendulums, each arm length is l and mass is m, are put on cars M. Angle from perpendicular axis to i-th pendulum is $\phi_i (i = 1, 2)$. Cars move according to external force F_i. In this model external force is given by van der Pol type. Interaction force f_i between pendulums is weighted on i-th pendulum. f_i is chosen to be proportional to angular difference between two pendulums. The proportional parameter d has important role in the phase transition. Our model is written as follows;

$$\alpha_i = \frac{F_i + mg \sin \phi_i \cos \phi_i - ml\dot{\phi}_i^2 \sin \phi_i}{M + m \sin^2 \phi_i},$$

$$\ddot{\phi}_i = \frac{1}{l}(g \sin \phi_i + \alpha_i \cos \phi_i + \frac{f_i}{m}),$$

$$F_i = -a_i \phi_i + b(\theta^2 - \phi_i^2)\dot{\phi}_i,$$

$$f_1 = -f_2 = d(\phi_2 - \phi_1), \qquad\qquad (i = 1, 2)$$

where α_i is acceleration of i-th car, g is acceleration of gravity and a_i, b, θ are constant parameters. Behavior of pendulums is given by using these equations.

3. Result

We define angular correlation A_ϕ as

$$A_\phi = |\phi_2 - \phi_1|,$$

which gives angular synchronization.

We calculate the motion of two pendulums by using fourth order Runge-Kutta algorithm. We choose constant parameters as follows;

$m = 0.1$, $M = 1.0$, $l = 0.5$, $a_1 = 16.4$, $a_2 = 16.3$, $b = 3.0$, $\theta = 0.1$.

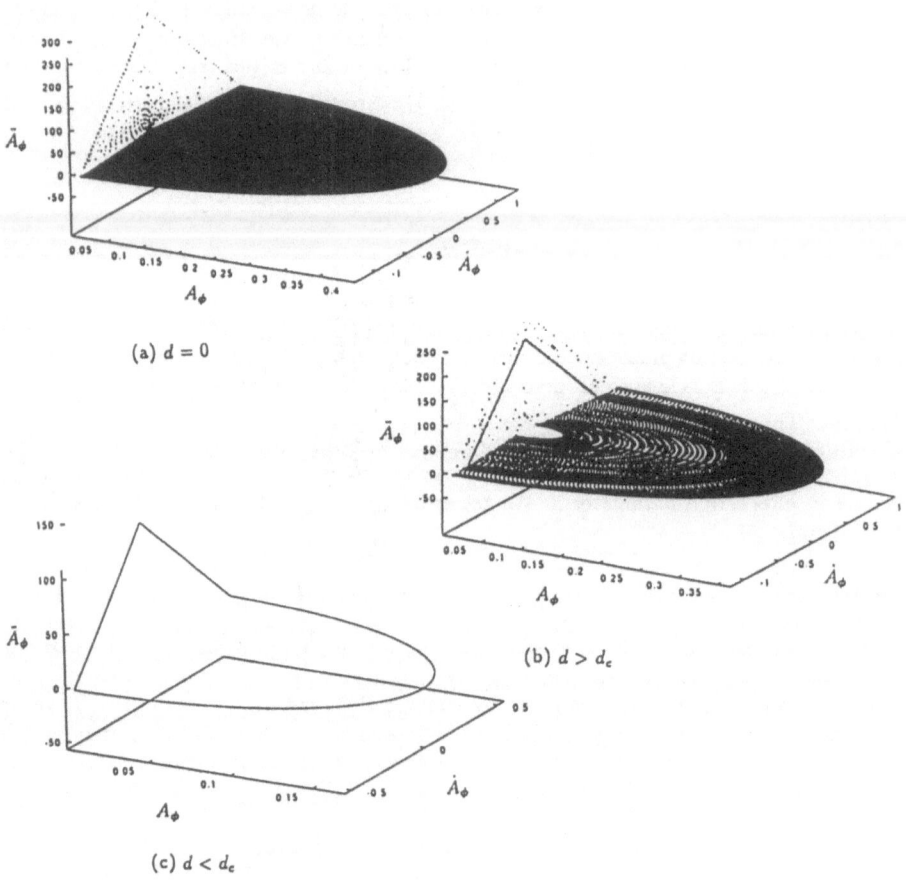

(a) $d = 0$

(b) $d > d_c$

(c) $d < d_c$

Fig. 1 Trajectory in $(A_\phi, \dot{A}_\phi, \ddot{A}_\phi)$ space for $d = 0$ (a), $d > d_c$ (b), $d < d_c$ (c).

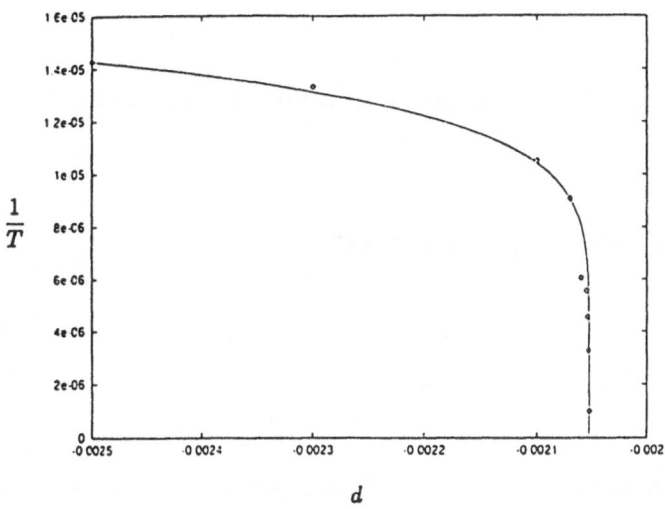

Fig. 2 Inverse convergent time $1/T$ versus interaction parameter d.

Fig. 1 shows trajectories in $(A_\phi, \dot{A}_\phi, \bar{A}_\phi)$ space for different coupling parameter d. In case of non-interaction $d = 0$, the trajectory is in the region of the oval as shown in Fig. 1 (a). Fig. 1 (b) shows the defect of the oval region for $0 > d > d_c$. In case of synchronization for $d < d_c$ (Fig. 1 (c)), the trajectory shows a curved line. We plot inverse convergent time $1/T$ versus the interaction parameter d in Fig. 2. We get critical point d_c and critical exponent δ using direct search of optimization;

$$\frac{1}{T} \sim (d_c - d)^\delta,$$

where we obtain $d_c = -0.002052 \pm 0.0000001$ and $\delta = 0.14 \pm 0.001$.

4. Discussion

In the presend paper we analyze the motion of two coupled van der Pol type inverted pendulums. We obtain critical point d_c and critical exponent δ in terms of angular correlation. Behavior of group of inverted pendulums will be invertigated in the base of this analysis.

References

[1] P. Bergé, Y. Pomeau and Ch. Vidal, *L'ordre dans le chaos*, Hermann, 1984
[2] I. Pastor-Díaz and A. López-Fraguas, Dynamics of coupled van der Pol oscillators, *Phys. Rev.* E **52**, 1995, pp. 1480-1489

Hamiltonian dynamics and statistics of relaxations

Yoshiyuki Y. Yamaguchi

Dept. of Physics, Nagoya Univ., Nagoya, 464-01, JAPAN

Abstract

Statistics of relaxations are investigated in a Hamiltonian system which has second order phase transition. Temporal evolutions of the system are yielded by Hamiltonian equations of motion, which are numerically integrated. Anomalously slow relaxation appears near the critical point for a statistical quantity. The statistic is produced by taking average over initial conditions.

Key words: Second order phase transition, critical point, slow relaxation, Hamiltonian dynamics, initial condition

1. Introduction

Relaxations to the equilibrium are interesting phenomena as dynamical properties of systems which have many degrees of freedom. In particular, we are interested in relaxations in second order phase transition, since the relaxation is anomalous near the critical point. That is the system comes closer to the critical point, the relaxation is the slower.

When we discuss relaxations of systems, we must get temporal evolutions of the systems. Systems having second order phase transition are represented by Hamiltonians, we therefore get temporal evolutions from Hamiltonian equations of motion, which are

$$\frac{dq_j}{dt} = \frac{\partial H}{\partial p_j}, \quad \frac{dp_j}{dt} = -\frac{\partial H}{\partial q_j}, \quad (j = 1, 2, \cdots, N) \tag{1}$$

where q_j and p_j are general coordinate and its canonical momentum respectively.

There are two advantages using Hamiltonian dynamics. First no assumptions are needed. When we use the other methods some assumptions are required as property of linear response for instance. Second we can directly obtain temporal evolutions of any quantities which are functions of q_j and p_j.

Some static and dynamic properties are reported for second order phase transition using Hamiltonian dynamics [1][2]. In [2] the present authour shows that slow relaxations of order parameter appears at the critical point for some initial conditions. However it is not obvious whether we observe anomalous relaxations even when we consider statistics about initial conditions, and which quantities show the anomalies. In this paper we introduce two statistic quantities, and reveal that one of them shows anomalous relaxation near the critical point.

This paper is constructed as follows. We introduce a model Hamiltonian and the observed quantities in Sec.2. Results of simulations are shown in Sec.3. Section 4 is devoted to summary.

2. Model and observed quantities

We introduce a system of classical spins, each of which is on a lattice point of a simple cubic lattice. The model Hamiltonian is represented as follows

$$H(q,p) = \frac{1}{2} \sum_{j=1}^{N} p_j^2 + \sum_{<jk>} (1 - \boldsymbol{m}_j \cdot \boldsymbol{m}_k), \quad N = L^3, \tag{2}$$

$$\boldsymbol{m}_j = (\cos q_j, \ \sin q_j), \quad q_j \in [0, 2\pi) \tag{3}$$

where j, k indicate lattice points of the cubic lattice, $<jk>$ means that summation is taken over all pairs of nearest neighbor lattice points and L is the size of each edge. We adopt periodic boundary condition.

Temporal evolutions of the system are yielded by Hamiltonian equations of motion

$$\frac{dq_j}{dt} = \frac{\partial H}{\partial p_j} = p_j, \quad \frac{dp_j}{dt} = -\frac{\partial H}{\partial q_j} = -\sum_{k|j} \sin(q_j - q_k) \tag{4}$$

where $j = 1, 2, \cdots, N$, and the summation of the second equation is taken over k which are nearest neighbor lattice points of j. We numerically integrate these equations of motion using symplectic integrator of fourth-order with a fixed time slice $\Delta t = 0.01$, which holds total energy accuracy as $|\Delta E/E| < 10^{-8}$. We initially take $q_j(0) = 0$ and $p_j(0) = \alpha x_j$, where x_j is produced from normalized Gaussian distribution, and α is determined from total energy.

We define a time-dependent order parameter of this system

$$M^i(t) = \frac{1}{N} \left| \sum_{j=1}^{N} m_j^i(t) \right|. \tag{5}$$

The time series of this order parameter depend on the initial conditions, which are distinguished by upper safix i. Next we take time-average of $M^i(t)$

$$\overline{M^i}(t) = \frac{1}{t} \int_0^t dt' M^i(t'). \tag{6}$$

As observed quantities we introduce average and deviation of $\overline{M^i}(t)$ over initial conditions

$$< \overline{M}(t) > = \frac{1}{S} \sum_{i=1}^{S} \overline{M^i}(t) \tag{7}$$

$$< (\Delta \overline{M}(t))^2 > = \frac{1}{S} \sum_{i=1}^{S} (\overline{M^i}(t) - < \overline{M}(t) >)^2, \tag{8}$$

where S is the number of initial conditions. The latter quantities represents fluctuation caused by initial conditions and how fast the system gets ergodicity because all of $\overline{M^i}(t)$ go to the unique ensemble average and $< (\Delta \overline{M}(t))^2 >$ equals zero when ergodicity holds [3].

3. Results of simulations

We show temporal evolutions of the two quantities, represented by eqs.(7) and (8) in Figs.1 and 2, respectively. Those figures includes results for $E/N = 3.0, 3.2$ and 3.5, where $E/N = 3.2$ is near the critical point. We find that $< \overline{M}(t) >$ shows systematic change as E/N increases. On the other hand, $< (\Delta \overline{M}(t))^2 >$ are anomalous near the critical point. Each temporal evolution of $< (\Delta \overline{M}(t))^2 >$ is approximated by a power function, such that $t^{-0.6}, t^{-0.4}$ and $t^{-1.4}$ for $E/N = 3.0, 3.2$ and 3.5 respectively. That is the realxation of $< (\Delta \overline{M}(t))^2 >$ is the slowest near the critical point.

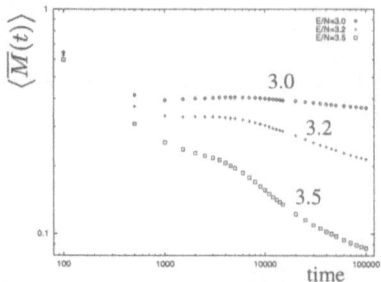

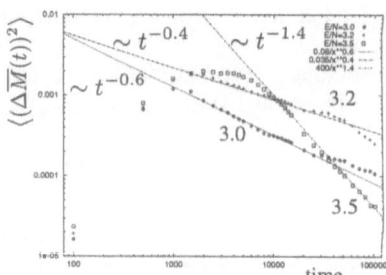

Fig.1: Temporal evolutions of $< \overline{M}(t) >$. Numbers in the figure represent E/N, where $N = 14^3$ and $E/N = 3.2$ is near the critical point. There is no anomaly even near the critical point.

Fig.2: Temporal evolutions of $< (\Delta \overline{M}(t))^2 >$. Numbers in the figure represents E/N, where $N = 14^3$. There is an anomaly near the critical point such that the relaxation is the slowest. Lines, which represent power funcionts, approximate $< (\Delta \overline{M}(t))^2 >$.

4. Summary

We consider slow relaxation of second order phase transition near the critical point. A model Hamiltonian is introduced, and we get temporal evolutions of the system by numerically integrating the Hamiltonian equations of motion. Relaxations of the system is observed through two quantities, which are average and deviation of time series of order parameter over initial conditions. The latter quantity shows anomalously slow relaxation near the critical point. That is, a system having second order phase transition shows an anomaly near the critical point even when we use Hamiltonian dynamics and take statics over initial conditions.

To understand the anomaly from motion in phase space is an interesting future work as a problem of classical Hamiltonian dynamics since, near the critical point, the motion is highly chaotic but not Markovian [2].

Acknowledgement

I thank to T. Konishi, H. Yamada, R. Akiyama for useful discussions.

References

[1] M. Antoni and S. Ruffo, Clustering and Relaxation in Hamiltonian Long-range Dynamics, *Phys. Rev. E* Vol.52, 1995, pp 2361-2374.

[2] Y. Y. Yamaguchi, Slow Relaxation at Critical Point of Second Order Phase Transition in a Highly Chaotic Hamiltonian System, *Prog. Theor. Phys.* Vol.95, 1996, pp 717-731.

[3] D. Thirumalai and R. D. Mountain, Ergodic Convergence Properties of Supercooled Liquids and Glasses, *Phys. Rev. A* Vol.42, 1990, pp 4574-4587, and references therein.

Part III Physical Systems

Complexity in a Molecular String: Hierarchical Structure as Is Exemplified in a DNA Chain

Kenichi Yoshikawa

Graduate School of Human Informatics, Nagoya Univ., Nagoya 464-01, Japan

Abstract

A long molecular string, or a polymer, undergoes a discrete transition between an elongated coil and a collapsed globule. Single-chain observations of natural long DNA chains have provided clear evidence that individual chains undergo a first-order phase transition. In contrast, the ensemble average of the chains appears to be continuous; *i.e.*, there is neither a first-order nor a second-order phase transition for the ensemble. This marked difference in the behavior of the transition between different levels in the ensemble is expected to be a general characteristic of hierarchical systems that consist of microscopic, mesoscopic and macroscopic levels.

Key Words: Statistical physics with a hierarchy, Self-organized nano-structure, Coil-globule transition, Crystallization of a single polymer, Collapse of DNA

§1. Introduction

In living cells, including both bacteria and eukaryote cells, long DNAs are packed within a small space without becoming entangled, and are hundreds of times more compact than when free in aqueous solution [1]. Considerable effort has been devoted to understanding the packing of long DNAs *in vivo* and *in vitro*. It is possible that the packing of DNA is involved in self-regulation of the metabolic state, and also in cell differentiation. According to the central dogma, the information stored in the sequence of bases along the DNA chain is transformed into proteins based on the code stored in RNA. The process of information transaction from DNA to protein molecules is called "gene expression." The dynamic process of gene expression has attracted the interest of biological scientists as a key phenomenon in living organisms. The most important unsolved problem concerning this phenomenon may be the mechanism of self-regulation of gene expression. Since the process of gene expression is driven by the dissipation of chemical energy within living cells, studies on the mechanism of self-regulation as it relates to nonlinear dynamics under nonequilibrium conditions may be quite important.

It has been believed that modulation of the higher-order structure of DNA must accompany all processes which involve reversible DNA compaction, such as encapsulation of DNA into reassembled viral proheads, the organization of chromatin in sperm cells, spores, and eukaryotic chromosomes, as well as replication and transcription. However, the factors that control and modulate the higher-order conformational modifications for such processes are far from being fully understood. Here, let us compare the mechanism of information processing in living organisms with that in modern computers. Both systems maintain their activity and function accompanied by a dissipation of free energy. It is well known that computers work well with both ROM, read only memory, and RAM, random access memory. In living organisms, the base-sequence in DNA corresponds to ROM. We propose the hypothesis that, in living cells, modulations in the higher-order structure of DNA may have a function similar to RAM [2,3]. A deeper understanding of the higher-order structure of DNA in an aqueous environment is expected to provide novel insight into the mechanism of self-regulation in gene,expression.

§2. Phase Transition in a String

First, we would like to present experimental evidence regarding the large discrete transition in individual single DNA chains. Figure 1(a) shows fluorescence images of T4DNA molecules in the coil (upper) and globule (lower) states. Coiled DNA was observed in aqueous solution and globular DNA was seen in 180μM spermidine (3+), $NH_3^+(CH_2)_3NH_2^+(CH_2)_4$ NH_3^+, solution. Figure 1(b) shows the spatial distribution of the fluorescence light intensity for the coil and globule DNAs. Figure 1(c) shows a schematic representation of the relationship between the fluorescence image and the microscopic structure of the DNA chain. As depicted in Fig. 1(c), a blurring effect occurs on the order of 0.3μm, due to the resolution limit attributed to the wavelength of the observation light and to the technical characteristics of the SIT camera [2-12]. Based on measurements of the diffusion constant for individual fluorescent images, the gyration radii of the coil and globule were estimated to be 1.6μm and 0.2μm, respectively. This implies that the effective volume changes *ca*. 500-fold with the coil-globule transition.

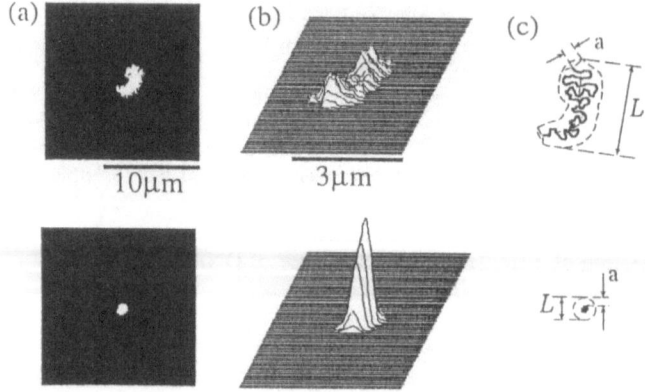

Fig. 1 [9] (a) Fluorescence microscopic images of T4DNA in the coil and globule states. We used Tris buffer solution (10mM Tris,1mM NaCl, at pH. 5.2), with 0.1μM, 4',6-diamidino-2-phenylindole (DAPI), and 4% (v/v) 2-mercaptoethanol.
(b) Spatial distribution of the fluorescence light intensity in (a). Due to the high density of DNA segments in the globule state, the fluorescence intensity is rather strong for globular DNAs.
(c) Schematic representation of the relationship between the conformation of an actual DNA chain and the corresponding fluorescence image. Due to the blurring effect ($a \approx 0.3$μm), the fluorescence image is larger than the actual DNA chain.

To characterize the structural transition of DNA induced by spermidine in a quantitative manner, a series of measurements was carried out by changing the spermidine concentration. As a measure of the effective size, the long-axis length L of DNA chains (Fig. 1(c)) was evaluated directly from the video images. The results are summarized in Fig. 2(a), which shows two maxima with spermidine concentrations of 140 and 160μM. In the actual observation, we can clearly distinguish between the coil and globule states [4], since the globule does not exhibit any visible conformational fluctuation, whereas the coil exhibits intramolecular Brownian motion. In Fig. 2(b), the solid circles show the maximum values of the long-axis length L. The shaded region in the figure corresponds to the bimodality in the distribution. The broken line corresponds to the transition profile of the ensemble average of the DNA chains. These results clearly show that **the transition between the coil and globule states is discrete for individual DNA chains, in the sense that each macromolecule assumes either the coil or globule state. On the other hand, the transition behavior for the ensemble average of the DNA chains is sigmoidal, which is typical for a cooperative and continuous transition.**

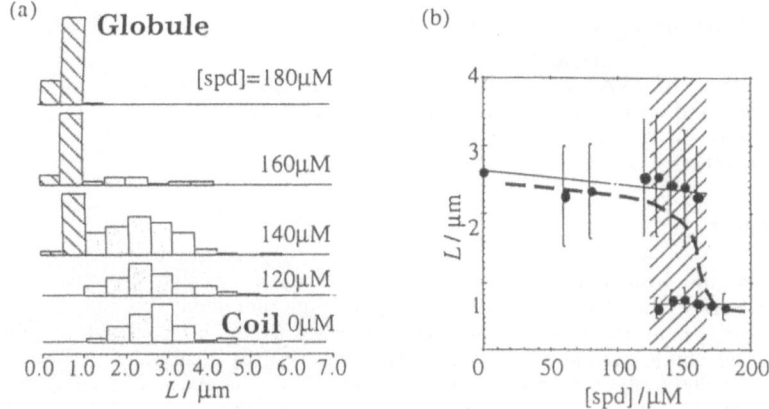

Fig. 2 [9] (a) Experimental histogram of the distribution of the long-axis lengths of T4DNA molecules at various concentrations of spermidine [spd]. The number of T4DNA's analyzed was *ca.*100. [DNA] = 0.10μM with the concentration expressed in base pairs. Each area of the histogram is normalized to be equal.

(b) Long-axis lengths of T4DNA molecules *vs.* the concentration of spermidine. The solid circles indicate the maxima for the coil and globule states, respectively, in the distribution of DNA lengths. The statistical width in the distribution is given as the standard deviation. The broken line represents the transition curve for the ensemble average of the long-axis lengths. The actual length, or the diameter, in the globule state is evaluated to be *ca.*0.4μm based on a measurement of the diffusion constant.

This result indicates that single polymer chains, such as DNA and other stiff polymers, may exhibit a first-order phase transition, even when the transition appears to be continuous in the ensemble. A similar discrete character of the transition between elongated coil and compacted globule has been observed for the collapse induced by a neutral hydrophilic polymer (polyethylene glycol) [4,8], a cationic surfactant [5,6] and a multivalent metal cation [10]. Let us consider a polymer chain which contains N monomer links with persistence length l and width d. The free energy F can be written as [9,13]:

$$\frac{F}{T} = \frac{3}{2}\left(\frac{1}{\alpha^2} + \alpha^2\right) + \frac{BN^{\frac{1}{2}}}{\alpha^3 l^3} + \frac{C}{\alpha^6 l^3} \qquad (1)$$

where the first two terms describe the elastic free energy, and the third and fourth terms are the free energy of interaction. B and C are the second and third virial coefficients for the interaction of the monomer links. The dependence of the equilibrium swelling coefficient α of the polymer chain on the solvent can be found by minimizing free energy F over α. As a result, we obtain the following equation:

$$\alpha^5 - \alpha - \frac{2y}{\alpha^3} = x\frac{d}{l} \qquad (2)$$

The parameter x characterizes the solvent quality: $x = B\,N^{1/2}/l^{1/2}d$. The parameter y $= C / l^6$ is related to the flexibility of the chain: the stiffer the macromolecule, the lower the value of parameter y. In Fig. 3(b) we plot typical theoretical dependencies of the swelling coefficient α of macromolecules undergoing a coil-globule transition as a function of parameter x, which represents the quality of the solvent. For chains that are sufficiently stiff ($l/d \gg 1$), the coil and globule states coexist in a region with a finite parameter width ($x_1 > x > x_2$). In Fig. 3(a), the size distribution function $p(\alpha)$ for a stiff polymer is given, where

$$p(\alpha) = \alpha^2 \exp[-\Delta F(\alpha)/T]/\int \alpha^2 \exp[-\Delta F(\alpha)/T] d\alpha \qquad (3)$$

On the other hand, if DNA collapse is observed by the usual experimental methods, such as light scattering and sedimentation, only the characteristics of the ensemble average are observed (see refs. 14-16 and 2,4-7 9-10). The mean swelling coefficient in the ensemble, $\langle\alpha\rangle$, is determined as follows:

$$\langle\alpha\rangle = \int_0^\infty \alpha p(\alpha) d\alpha \qquad (4)$$

In such cases, the broken curve in Fig. 3(b) is observed, which is consistent with the experimental results with DNA (Fig. 2(b)). Thus, although individual stiff chains should show a discrete first-order conformation transition (as manifested by the bimodal size distribution), the results on the ensemble average which are actually observed by standard experimental methods show continuous sigmoidal curves with a finite width. Since the above theoretical consideration is rather general, this unique characteristic of polymer chains should also apply for other natural and synthetic "stiff polymers."

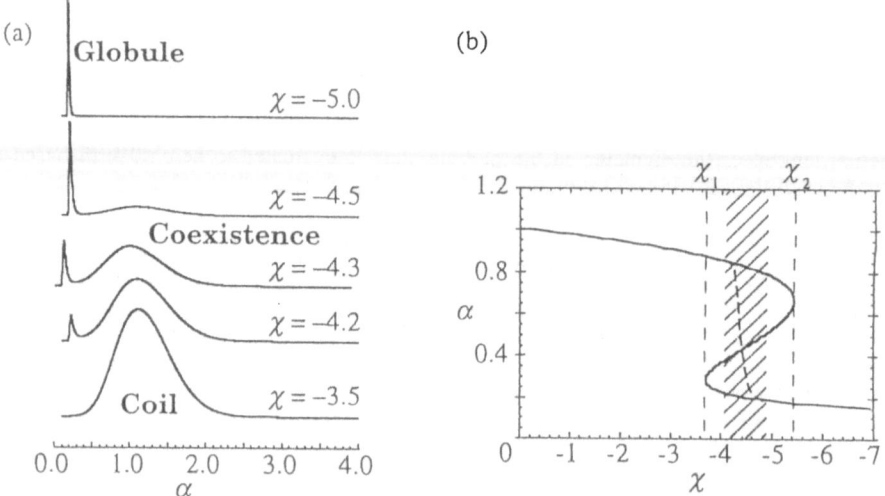

Fig.3 [9]   (a) Theoretical size distribution for the ensemble of a stiff polymer, $y = 10^{-3}$, depending on the solvent quality.
(b) Unique transition characteristics of a stiff polymer. For a single chain, the bistable property is represented by the Z-shaped solid line. For an ensemble of chains, the transition appears to be continuous, as shown by the broken line. The shaded part indicates the coexistence region of coil and globule DNAs.

§3. Phase Segregation in a String

In the preceding section, we clearly demonstrated the first-order nature of the phase transition in a single DNA chain, both experimentally and theoretically. The all-or-none nature of the transition implies phase-separation between different chains. It is expected that phase-separation should occur along a single chain under appropriate conditions. In this section we discuss the different types of phase separation, intrachain or interchain, in relation to the change in free energy in a single chain.

In general, the size R of a linear chain varies as $R \sim N^{1/2}$ in the ideal state and $R \sim N^{3/5}$ in the actual state. It has been suggested that the chain size varies as $N^{1/2}$ for a short range, even in a real chain. The fractal dimension of a chain depends on the resolution; for low resolution, $d = 5 / 3$; for high resolution, $d = 2$ [13, 17-18]. The cross-over between the two different scaling laws occurs at a certain distance: a long polymer chain is regarded as a successive linear array of thermal "blobs," as in Fig. 4.

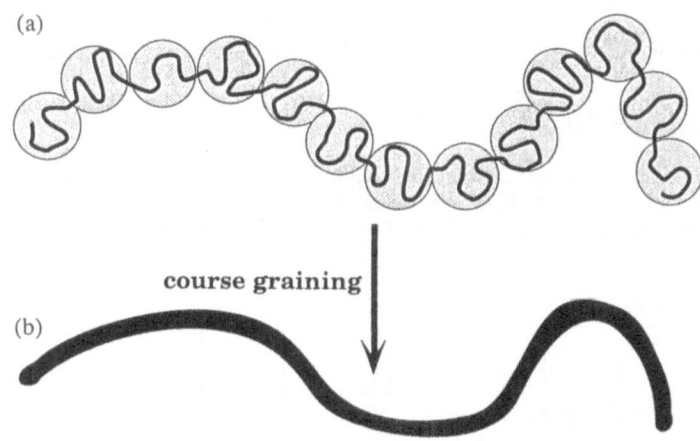

(a)

course graining

(b)

apparent contour length: L

Fig. 4 (a) Long polymer chain considered as a strand of connected blobs.
(b) With course-graining, the polymer chain looks like a smooth chain, where the apparent contour length L is much shorter than the full stretched length of the polymer chain, as in (a).

Let us consider the free energy per blob to be $g(\eta)$, where η is the dimensionless density (order parameter), $\eta = (\rho - \rho_c) / \rho_c$, and ρ and ρ_c are the densities of the segments within a blob in an actual chain and in the coil state with the free energy minimum, respectively. If the free energy of the blob is expressed with double minima with respect to η, as is usual in a first-order phase transition, $g(\eta)$ can be written as follows [19]:

$$g(\eta) = g_0 + a(T - T_0)\eta^2 + b\eta^4 + c\eta^6 + A(\nabla\eta)^2 \qquad (5)$$

where $a > 0$, $b < 0$, $c > 0$, and $A > 0$. If we define the phase transition temperature $T = T_c$ as that at which $g(\eta)$ has the same value at the minimum of the coil state ($\eta = 0$) and at the minimum of the globule state ($\eta = \eta_g$), $g(\eta)$ should be

$$g(\eta) = c\eta^2\left(\eta^2 - \eta_g^2\right)^2 + A(\nabla\eta)^2 \qquad (6)$$

Comparing equations (5) and (6),

$$\eta_g^4 = a(T_c - T_0)/c \qquad (7)$$

$$b = -2c\eta_g^2 \qquad (8)$$

From equations (7) and (8),

$$T_c - T_0 = \frac{b^2}{4ac} \tag{9}$$

Using the dissipation-fluctuation theory, the "susceptibility" of η at temperature T is given by

$$\chi(T) = \frac{1}{2\left[Ak^2 + a(T - T_0)\right]} = \frac{1}{2a\left(1 + k^2\xi(T)^2\right)(T - T_0)} \tag{10}$$

where the correlation length $\xi(T) = [A / a (T - T_0)]^{1/2}$. Thus, at the critical temperature,

$$\xi_c = \left[A/a(T_c - T_0)\right]^{1/2} = 2(Ac)^{1/2}/|b| \tag{11}$$

Equation (11) indicates that the correlation length ξ increases when c increases. It is noted from eq. (6) that the free energy barrier between the coil and globule states increases with the increase of c. Thus, the dependence of ξ_c on the parameters given in eq. (5) becomes clear.

(a) $L < \xi_c$

(b) $L > \xi_c$

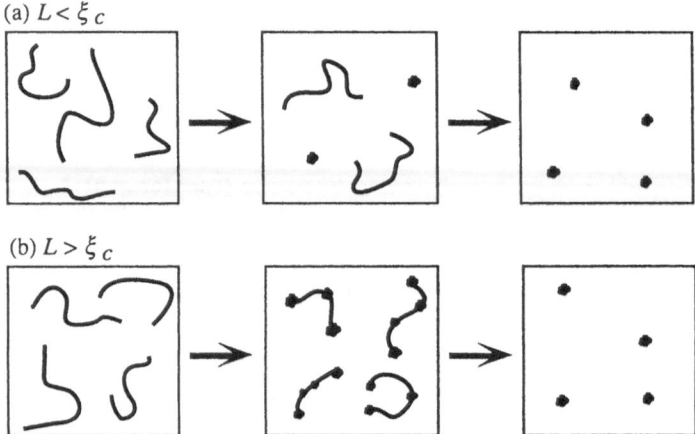

Fig. 5 Two different routes for the coil-globule change through a first-order phase transition. (a) the correlation length ξ_c is longer than the system size (L: the size of a polymer chain, see Fig. 4), (b) the correlation length ξ_c is smaller than the system size.

This theoretical consideration leads to the following prediction regarding the manner of the transition. If the contour length L in the course-grained chain (c.f., Fig. 4) is smaller than ξ_c, i.e., $L < \xi_c$, an all-or-none transition is induced for a single chain, as is shown in Fig. 5(a). On the other hand, if $L > \xi_c$, intrachain segregation will be induced, i.e., coexistence of the elongated and collapsed states along a single chain (see Fig. 5(b)).

Based on the results and discussion in §2 and §3, the apparent features of the transition for an ensemble of chains can be classified as in Fig. 6. Even in the case of infinite numbers of polymer segments, N, and polymer chains, M; (i.e., both $N \rightarrow +\infty$ and $M \rightarrow +\infty$), the width of the transition remains a finite value. Thus, the transition at the "thermodynamic limit" is not discrete. It may be useful to compare with the transition profile for the ensemble of flexible polymer chains. In this case, the width is expected to vary as $\Delta T \sim N^{1/2}$[13], regardless of M. Such a characteristics of the phase transition in polymers seems to be general for "mesoscopic systems." Recently, similar theoretical discussion has been reported on the statistical characteristics of clusters composed with finite number of atoms[21,22].

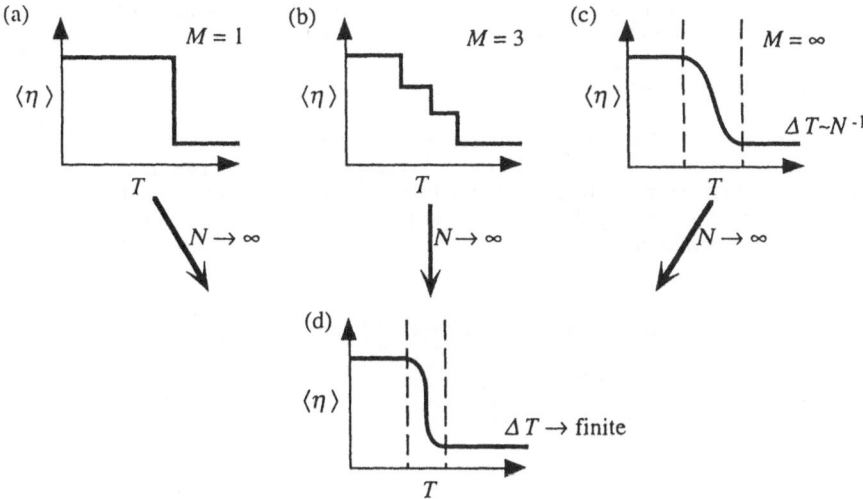

Fig. 6   Schematic representation of the change in the order parameter $\langle \eta \rangle$ (average density) of polymer chain(s). N and M are the numbers of segments and polymer chains, respectively. N is finite in (a), (b), and (c), and infinite in (d). Due to intrachain segregation, the width of the transition remains a finite value, even in the infinite limit of the chain length, which is related to the finiteness on the correlation length, ξ_c, in first order phase transition (eq.(11)).

§4. Crystallization in a String

It is now clear that a long single string exhibits a first-order phase transition, as exemplified by the experimental results with giant DNA. Since this transition is first-order, a single chain is expected to exhibit the kinetics of "nucleation and growth." In this section, we present an actual example of the observation of nucleation and growth in a single DNA chain. We also discuss theoretically the kinetic process of crystallization in a single polymer chain.

Figure 7 shows the forward and reverse transition of a single double-stranded T4DNA chain from an elongated random coil into a compacted globule in an aqueous solution of PEG [23]. This transition is induced by the addition of the neutral hydrophilic polymer PEG. The region of coexistence of the coil and globule states is found between 5.5 and 6.5M of PEG under these experimental conditions. Thus, at [PEG] = 6.8M, as shown on the left in Fig. 2, the individual coil is metastable and has a lifetime on the order of several tens of minutes. During Brownian motion of the metastable coil, a bright spot appears spontaneously at a certain point on the chain. As the fluorescence intensity at the brightest spot gradually increases, the apparent contour length of the chain decreases. The process of compaction from coil to globule is completed within 10s.

The time course of this compaction was analyzed by measuring the apparent contour length, L, and the light intensity at the brightest spot. Figure 7 shows that both L and the fluorescence intensity exhibit linear dependence (as for the detail of the kinetic data, see ref. 23). Thus, the process of compaction from a metastable coil to a globule is similar to nucleation and growth from a metastable or supersaturated fluid state to a highly condensed regular state or crystal state. The essential difference between the time course of the coil-globule transition of a single DNA chain as a linear string and the "nucleation and growth" observed in the usual first-order transition is the dimensionality of the system. The constancy of the rate of growth is explained by considering the unique aspects of the "nucleation and growth" of a single chain[2], *i.e.*, of the remaining coiled portion: only that which is immediately adjacent to the globule is pulled into the condensed globular "crystal."

The time profile of the reverse transition from a globule into a coil is shown in Fig. 7(a)'- (e)'. In contrast to the linearity of the transition from a metastable coil to a globule with respect to time, the reverse transition does not proceed at a constant speed; *i.e.*, the process is initially slow, and then gradually accelerates[23]. This asymmetry between the collapsing and decollapsing transitions is interesting in relation to a problem concerning the folding of proteins: how does a protein reach the global free energy minimum from among a tremendously large number of local energy minima?

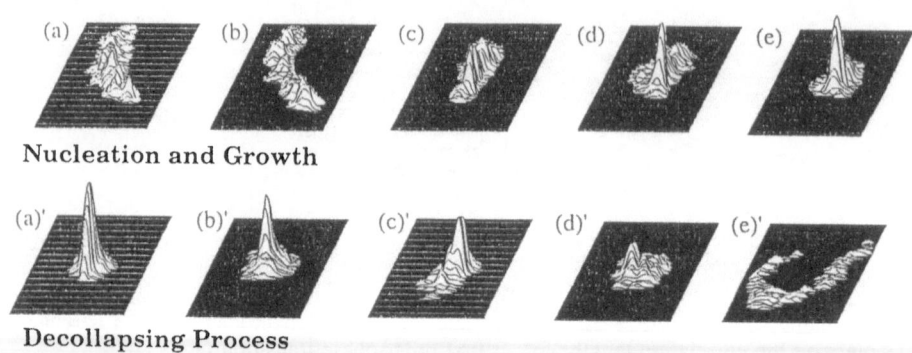

Nucleation and Growth

Decollapsing Process

Fig. 7 [22] Dynamic process of the transition of a single T4 DNA molecule. The height in the quasi-three-dimensional representation indicates the intensity of the fluorescence, which corresponds to the spatial density of the segments in a single DNA. We confirmed that the phenomenon of nucleation and growth is general, by observing this process for *ca.* 50 molecules. In these observations, nucleation was most frequently initiated at the end of the DNA chain[11].
Upper: Transition from an elongated coil into a compacted globule. The kinetic process is characterized as nucleation and growth. The time interval is 2 sec.
Lower: Transition from a globule into a coil. The time interval is 3 sec., except for the 20-sec period between d)' and e)'.

(a)   (b)   (c)

toroidal structure   complex structure   rod structure

Fig. 8 [24] Self-organized structures of elongated coiled polymer chains obtained by Monte-Carlo simulation.

Figure 8 shows the results of the Monte-Carlo simulation [24] of collapsed products from a single polymer chain without electronic charge. For toroidal structures (Fig.8 (a)), the average diameter corresponds to 600 $\mathring{A}$, when the persistence length in the simulation corresponds to *ca.*500 $\mathring{A}$, as in the case of a DNA chain. To compare the results of the simulation with actual experimental results, the structures of single T4 DNAs compacted with hexammine cobalt (III) are given in Fig. 9. In this experiment, the inner and outer diameters were 325 and 820Å, respectively, for the ensemble average in the toroids. Thus, the mean diameter is *ca*.570 Å. Therefore, the results of the computer simulation apparently reproduce the experimentally observed toroidal structure in a semi-quantitative manner. In addition, the simulation also reproduces the other experimental morphologies fairly well.

(a)   (b)   (c)

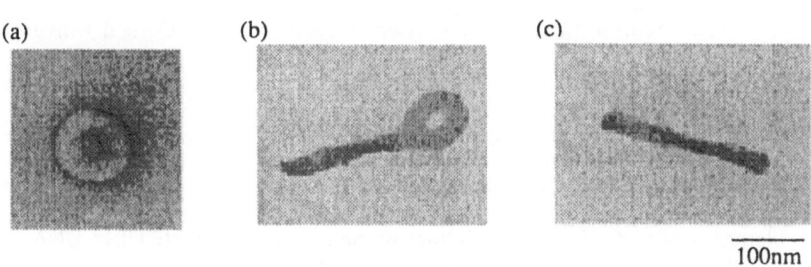

100nm

Fig. 9 [24] Transmission electron microscopy of T4 DNA compacted by hexammine cobalt (III).

References

[1] V. A. Bloomfield, DNA Condensation, *Curr. Opin. Struct. Biol.* Vol. 6, 1996, pp. 334-341.

[2] K. Yoshikawa, Y. Matsuzawa, Discrete Phase Transition of Giant DNA. Dynamics of Globule Formation from a Single Molecular Chain, *Physica*, Vol. D84, 1995, pp. 220-237.

[3] S. Kidoaki, K. Yoshikawa, The Folded State of Long Duplex-DNA Chain Reflects Its Solution History, *Biophys. J.*, Vol. 71, 1996, pp. 932-939.

[4] V. V. Vasilevskaya, A.R. Khokhlov, Y. Matsuzawa, K. Yoshikawa, Collapse of Single DNA Molecule in Poly(ethylene glycol) Solutions, *J. Chem. Phys.*, Vol. 102, 1995, pp. 6595-6602.

[5] S. M. Mel'nikov, V. G. Sergeyev, K. Yoshikawa, Discrete Coil-Globule Transition of Large DNA Induced by Cationic Surfactant, *J. Am. Chem. Soc.*, Vol.117, 1995, pp. 2401-2408.

[6] S. M. Mel'nikov, V. G. Sergeyev, K. Yoshikawa, Transition of Double-Stranded DNA Chains between Random Coil and Compact Globule States Induced by Cooperative Binding of Cationic Surfactant, *J. Am. Chem. Soc.*, Vol.117, 1995, pp. 9951-9956.

[7] Y. Yoshikawa, K. Yoshikawa, Diaminoalkanes with an Odd Number of Carbon Atoms Induce Compaction of a single Double-Stranded DNA Chain, *FEBS Lett.*, Vol. 361, 1995, pp. 277-281.

[8] K. Minagawa, Y. Matsuzawa, K. Yoshikawa, A. R. Khokhlov, M. Doi, Direct Observation of the Coil-globule Transition in DNA Molecules, *Biopolymers*, Vol. 34, 1994, pp. 555-558.

[9] K. Yoshikawa, M. Takahashi, V. V. Vasilevskaya, A. R. Khokhlov, Large Discrete Transition in a Single DNA Molecule Appears Continuous in the Ensemble, *Phys. Rev. Lett.*, Vol. 76, 1996, pp. 3029-3031.

[10] K. Yoshikawa, S. Kidoaki, V. V. Vasilevskaya, A. R. Khokhlov, Marked Discreteness on the Coil-Globule Transition of Single Duplex DNA, *Ber. Bunsenges. Phys. Chem.*, Vol. 100, 1996, pp. 876-880.

[11] Y. Matsuzawa, Y. Yonezawa, K. Yoshikawa, Formation of Nucleation Center in Single Double-stranded DNA Chain, *Biochem. Biophys. Res. Commun.*, Vol. 225, 1996, pp. 796-800.

[12] Y. Yoshikawa, N. Emi, T. Kanbe, K. Yoshikawa, H. Saito, Folding and Aggregation of DNA Chains Induced by Complexation with Lipospermine: Formation of a Nucleosome-like Structure and Network Assembly, *FEBS Lett.*, Vol. 396, 1996, pp. 71-76.

[13] A. Yu. Grosberg, A. R. Khokhlov. *Statistical Physics of Macromolecules*, American Institute of Physics, New York, 1994.

[14] C. B. Post, B. H. Zimm, Theory of DNA Condensation: Collapse Versus Aggregation, *Biopolymers*, Vol. 21, 1982, pp. 2123-2137.

[15] H. Yamakawa, The Radius Expansion Factor and Second Virial Coefficient for Polymer Chains below the Θ Temperature, *Macromolecules*, Vol. 26, 1993, pp. 5061-5066.

[16] T. Norisue, Y. Nakamura, Remarks on Smoothed-density Theories for Flexible Chains with Three-segment Interactions, *Polymer*, Vol. 34, 1993, pp. 1440-1443.

[17] J-L. Barrat, J-F. Joanny, Theory of Polyelectrolyte Solutions, *Polymeric Systems*, I. Prigogine and S. A. Rice eds., 1996, pp. 1-66.

[18] P-G. de Gennes. *Scaling Concepts in Polymer Physics, 2nd ed.,* Cornell University Press, Ithaca and London, 1993.

[19] L. D. Landau, E. M. Lifshitz. *Statistical Physics, 3rd ed., Part 1*, Engl. Trans., Pergamon, 1996.

[20] J. Jellinek, T. L. Beck, R. S. Berry, Solid-Liquid Phase Chantes in Simulated Isoenergetic Ar_{13}, *J. Chem. Phys.*, Vol. 84, 1986, pp.2783-2794.

[21] D. J. Wales, J. P. K. Doye, Coexistence and Phase Separation in Clusters: From the Small to the Not-So-Small Regime, *J. Chem. Phys.*, Vol. 103, 1995, pp. 3061-3070.

[22] S. G. Starodoubtsev, K. Yoshikawa, Intrachain Setregation in Single Giant DNA Molecules Induced by Poly(2-vinylpyrrolidone), *J. Phys. Chem.*, in press.

[23] K. Yoshikawa, Y. Matsuzawa, Nucleation and Growth in Single DNA Molecules, *J. Am . Chem. Soc.*, Vol. 118, 1996, pp.929-930.

[24] H. Noguchi, S. Saito, S. Kidoaki, K. Yoshikawa, Self-Organized Nanostructures Constructed with a Single Polymer Chain, *Chem. Phys. Lett.*, Vol. 261, 1996, pp. 527-533.

Fractal distributions - from the real world to the information world

Hideki Takayasu[1] , Misako Takayasu[2], and Takamitsu Sato[3]

1 Graduate School of Information Sciences, Tohoku University, Sendai 980-77, Japan
2 Center for Polymer Studies, Boston University, Boston, MA02215, USA
3 Multimedia Service Department, NTT, 2-2-3 Uchisaiwaicho, Chiyodaku, Tokyo 100, Japan

Abstract

Self-organization mechanisms of producing fractal distributions in the real world are reviewed showing several examples. Then we extend our subjects of research to the information world which are realized recently by the network of high performance computers. We find that basic quantities in the information world tend to follow fractal distributions, such as file sizes, elapsed CPU times for various jobs and density fluctuations of packet traffics.

Key words: Fractal distributions, power laws, file sizes, information traffics

1. Introduction

Nearly two decades have passed since the birth of the concept of "fractals" and it is now widely accepted that many complex phenomena in nature show fractal properties. In view of distribution fractal properties can be represented by power laws because they are the only scale-invariant distributions[1]. There is a strong mathematical theorem 'the generalized central limit theorem' which ensures the popularity of power law distributions[1,2]. According to this theorem a sum of many independent random variables having divergent variances is likely to follow a stable distribution which has a power law tail. In view of physics if we can decompose a phenomenon into superposition of elementary random processes whose variances are divergent, then we may expect a power law tail of a stable distribution to be observable.

Although the generalized central limit theorem looks very powerful, the requirement of independence is generally not satisfied in most of complex phenomena in the real world. In order to understand the mechanism of realizing power law distributions we need case studies as there is no general theorem for correlated random processes.

In the following section we review several examples in the real world. Topics to be reported are the sizes of fish schools and aerosols, river patterns created by water erosion and fragments of fragile materials. In the third section we look for power law distributions in a computer and in the largest computer network, the Internet. We show that the most basic distributions in the information world such as file sizes and CPU times show tendency to follow power laws. Information flows in the Internet also show fractal properties and we report recent results in the latter part of the same section. The last section is devoted for general discussion.

2. Examples in the real world

There are many examples of phenomena following power law distributions in nature. In order to understand the mechanism of realizing such power laws, Bak et al proposed a model of sandpile as an example of "self-organized criticality"[3]. Pietronero et al proved that the model actually converges in the macroscopic limit to the critical point of phase transition[4]. At the critical point of second order phase transition it is well-established that a characteristic length of correlation diverges and distribution of cluster sizes, for example, follows a power-law. The essential puzzle is the stabilization mechanism of the critical point since critical points are generally unstable in thermal equilibrium systems.

2.1 Fish schools

Many kinds of fish show tendency to make schools of various sizes. The distribution of fish school sizes can be observed through the distribution of mass of fish caught by a single catch with a large net. Real data of fish school sizes (to be denoted as m) demonstrate that the distribution is well approximated by a power law[5] predicted for so-called aggregation system with injection[6].

$$P(\geq m) \equiv \int_{m}^{\infty} p(m)\mathrm{d}m \propto m^{-0.5} \tag{1}$$

The aggregation system with injection can be regarded as a model of aerosols and we consider sticky random walkers which coagulate by collision. In the absence of external injection the number of clusters decrease monotonically by repeating coagulation and there is only one steady state in which all particles make a single cluster. By repeating injection of small mass particles with a constant rate a nontrivial statistically steady state realizes. Each cluster grows indefinitely, however, the number density of clusters and the mass distribution of clusters converge in the steady state. It is proven that the constant injection makes the critical point of phase transition stable and mass distribution in the steady-state generally follows a power law[7]. Aerosol cluster size distributions are actually approximated well by power law distributions[8].

A fish school can be viewed as a cluster, and if two clusters of fish of an identical species meet then we may assume that the two clusters make one cluster irreversibly. Also fish eggs may be scattered all over the ocean and it may be natural to conjecture that new fish are born with a constant rate. Therefore the fish school sizes follow the same statistics as aerosols.

2.2 Water erosion

Mandelbrot demonstrated that the shape of earth's surface can be modeled by fractional Brownian surfaces[9]. The outlook of both surfaces are actually similar, in the sense they are characterized by a fractional dimension, however, there is a big difference if we consider water flow over the surfaces. Namely, water does not flow out smoothly on a fractional Brownian surface, on the other hand water flows make river patterns on a real surface.

The effect of water erosion has been analyzed intensively by numerical simulations[10] and it is now evident that water erosion enhances small initial fluctuations on the surface and automatically create a fractal landscape. A fractal distribution can be found in basin sizes of a river network. Fig.1 shows an example of river patterns numerically created by a time evolution based on water erosion. For a given point on the river pattern a basin size is defined by the area of upper streams flowing to the point. Namely, in the case that rain falls constantly and uniformly the basin size is proportional to water flow intensity (to be denoted as n). In Fig.2 an example of basin size distribution is plotted. The cumulative distribution clearly follows a power law with exponent about 0.4, namely,

$$P(\geq n) \propto n^{-0.4} \tag{2}$$

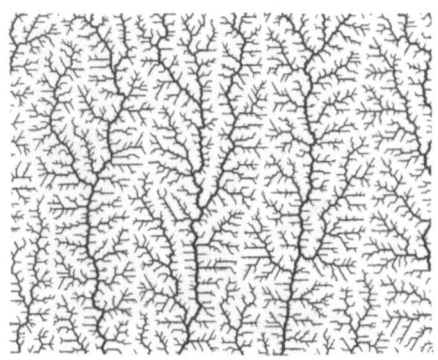

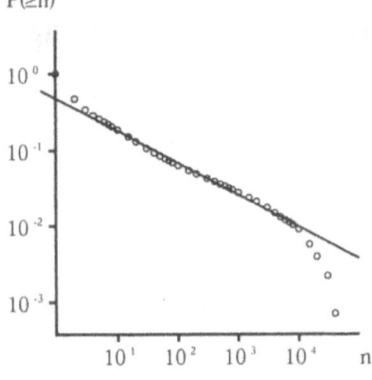

Fig. 1 Numerically created river pattern

Fig.2 River basin size distribution

This value of exponent is consistent with values for real rivers[11]. It is known that the self-organization mechanism of this fractal basin size distribution is so universal that the exponent value is independent of initial or boundary conditions[10].

2.3 Fragment sizes

When you drop a glass on the floor it breaks instantly into many pieces of various sizes. A clear empirical power law is known for the fragment pieces created by such impact fracture of brittle amorphous materials[12]. It is confirmed experimentally that the fragment mass follows a power law,

$$P(\geq s) \propto s^{-b} \quad , \quad b = 2/3 \tag{3}$$

Recently a numerical simulation of 3-dimensional impact fracture succeeded in deriving this power law[13]. The numerical model is based on the assumption that a density plain wave is created together with many small cracks on the impact surface. The cracks develop with the propagation of the density wave. In the case there are random fluctuations in the material the cracks on the density wave move randomly according to the fluctuations. When two or more cracks collide on the density wave they merge making one crack. The traces of cracks give a 3-dimensional crack configuration. Fig.3a shows an example of resulting crack pattern parallel to the impact surface plain and Fig.3b gives an intersection perpendicular to Fig.3a. A fragment mass is defined by the volume which is covered by cracks. In Fig.4 the mass distribution is plotted in log-log scale, and we can find a power law consistent with the empirical law, Eq.(2).

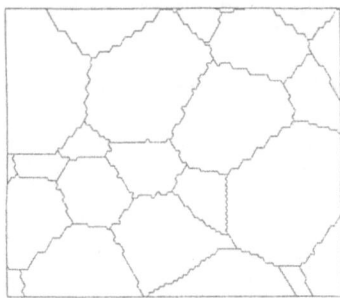

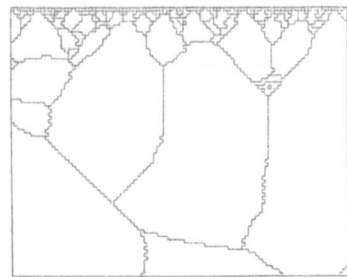

Fig.3a A cross-section of crack pattern
parallel to the impact surface

Fig.3b A cross-section of crack pattern
perpendicular to the impact surface

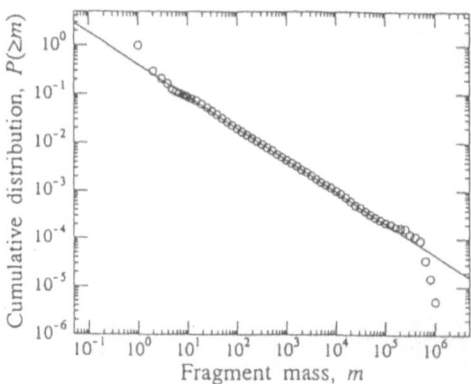

Fig.4 Numerically obtained fragment mass distribution

The simulation has also revealed the robustness of this exponent. By changing the details of crack dynamics the identical power law is obtained as far as the crack dynamics holds the property that a larger piece of fragment grows faster than neighboring smaller pieces. This result strongly supports the experimentally established universality of impact fracture phenomenon[13].

3. Examples in the information world

As we have seen in the previous section fractal distributions can be found in various fields of sciences. Roughly speaking it looks that a huge system having competitive growth mechanism of random fluctuations is likely to show fractal behaviors. Fractal properties can also be found inside of computers. Nowadays even a personal computer has a large information of order of 10^9 bytes. The Internet is the largest computer network consisted of computers all over the world. The information world supported by these computers has grown so big that we may expect fractal properties.

3.1 File or directory sizes

In the information world all data and programs are stored in the form of files, which are arranged in hierarchical directory structures. These structures of information are having tree-like branching forms topologically identical to the river patterns in the real world. In the case of information trees a directory corresponds to a branching point in rivers and a sum of file sizes belonging to a directory represents its "basin size". We observe basin size distributions in our laboratory and a typical cumulative distribution is plotted in Fig.5.

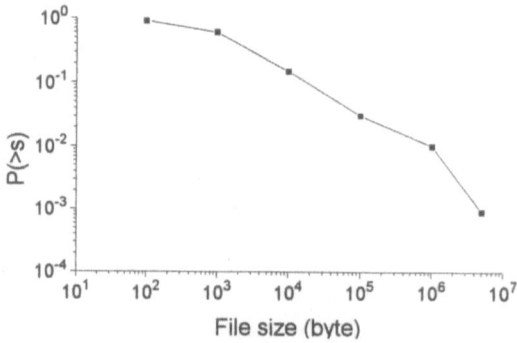

Fig.5 Basin size distribution in a hard-disk

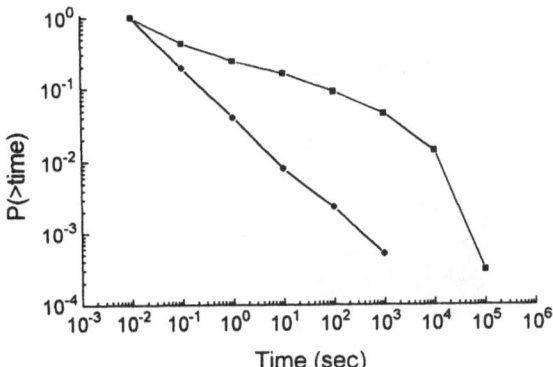

Fig.6 Distributions of CPU times (circles) and job existing times (squares)

We can find a long tail which is approximated by a power law with exponent roughly estimated to be 0.5. This exponent is the same as the cases of fish schools and randomly growing branching structures in a high dimensional space.

As far as we searched for several cases basin size distributions in hard-disks deviate clearly from a power law when hard-disks are nearly full. We may conjecture that the power law behavior of basin sizes in hard-disks is a result of free growth of files.

3.2 CPU times

Workstations in our offices are working day and night processing jobs of many kinds like scientific calculations, graphical data constructions and file transfers. What is the size distribution of jobs?

A size of a job can be estimated by the CPU time needed to finish the job. In Fig.6 the cumulative distribution of net CPU times for nearly 10^4 jobs is plotted. A clear power law with exponent about 0.7 can be found.

Another way of estimation of load for a job is given by a job existing time which is the gross time needed to finish the job. The distribution of job existing times also follows a power law with a different exponent (also plotted in Fig.6). The exponent is estimated to be about 0.3 which is nearly half of the exponent for the CPU times.

The relation between these two distributions may be due to a scaling relation between the CPU time (denoted by t) and the existing time (denoted by T) like $T \propto t^a$, a = 2 . However, no theoretical explanation has been established for these distributions and for this scaling relation.

3.3 Traffic fluctuation

In the Internet all data and messages are transferred in the form of packets. Recently it was found that the flow density fluctuation of packets observed at a point of the network shows a fractal property[14]. Fig.7a is an example of temporal fluctuation of packets observed at the gateway of Tohoku university. Fig.7b and Fig.7c are enlarged parts of Fig.7a with unit times 10 and 100 times smaller, respectively. Both enlarged figures look similar to the original demonstrating the self-similarity. As a consequence of the self-similarity the power spectrum of such traffic density fluctuation follows the 1/f power spectrum[15].

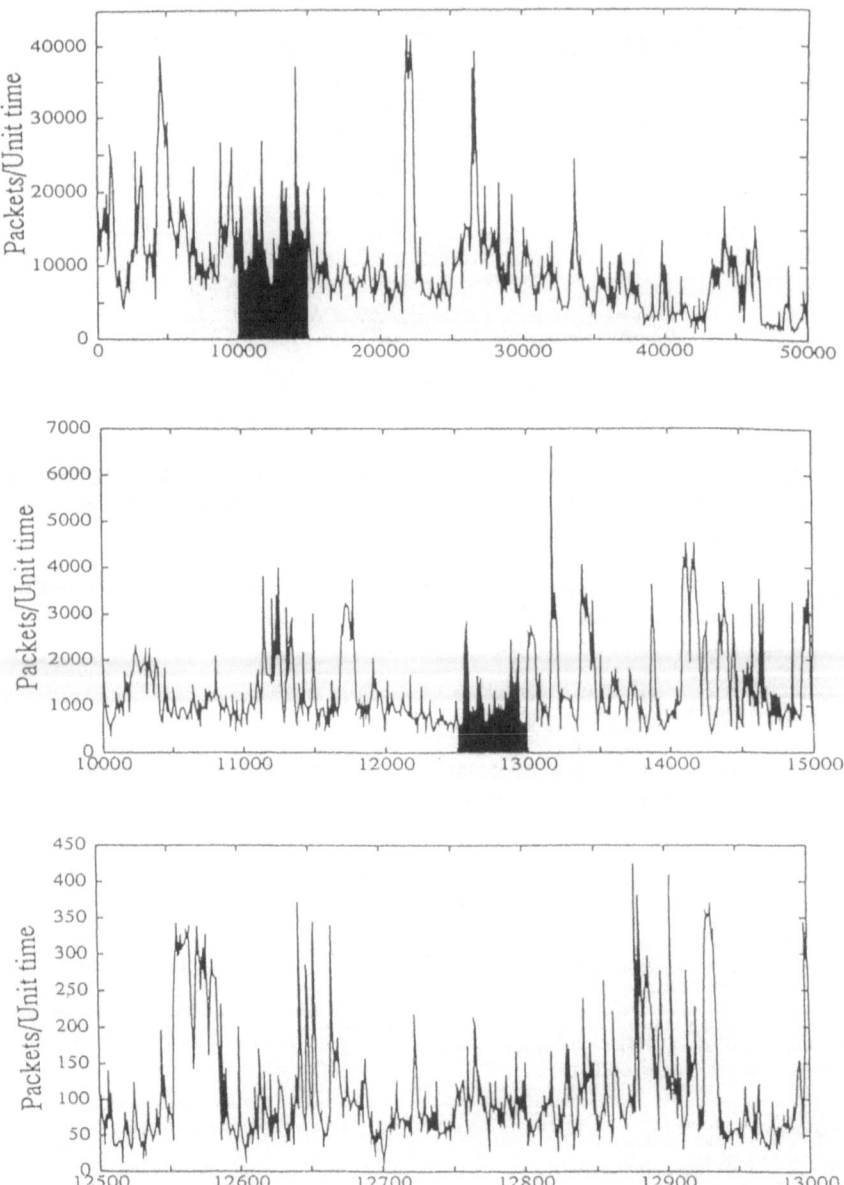

Fig.7 Fluctuation of packet flow density at a gateway
Time units are 100sec.(top), 10sec.(middle), 1sec.(bottom)

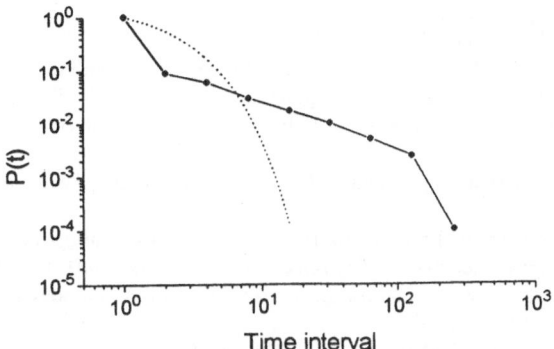

Fig.8 Interval distribution of level set for traffic density fluctuation (bold line)
A theoretical Poisson distribution having the same mean value (dotted line)

A power law distribution hidden in the traffic fluctuation becomes observable by using the level set method[16]. We set a threshold density and observe the time interval of duration of low density states. Fig.8 shows a cumulative distribution of the intervals in log-log scale. The distribution deviates clearly from the theoretical Poisson distribution which gives the observed mean value (dotted line in Fig.8), but it is much closer to a power law, $P(\geq t) \propto t^{-1}$. This interval distribution is consistent with the 1/f power spectrum[15].

The mechanism of producing this fractal fluctuation has been studied intensively. It is conjectured and confirmed partly by "Ping experiment" that jamming of packets at a router can propagate to neighboring routers and such interaction on a branched network causes a global dynamic phase transition between jamming and non-jamming states[15]. In this phase transition the control parameter is the mean packet injection rate and the information flow is most efficient at the critical point. Theoretical analysis revealed that the branched structure of the network is essential for the 1/f behavior.

4. Concluding remarks

Fractal distributions have been found not only in the real world but also in the information world created in computers as we have seen. Therefore, the most essential mechanism of realizing such distributions should not depend on material dynamics. A very universal mathematical mechanism effective in both worlds may be underlying among these fractal phenomena.

References
[1] Hideki Takayasu. *Fractals in the physical sciences*, Manchester University Press, Manchester, 1990.
[2] W.Feller. *An introduction to probability theory and its applications*, Wiley, New York, 1966.
[3]P.Bak, K.Chen, Self-organized criticality, *Sci. Amer.*, vol.264, 1991, pp.46-53.
[4] L.Pietronero, W.R.Schneider, Fixed scale transformation approach to the nature of relaxation clusters in self-organized criticality, *Phys.Rev.Lett.*, vol.66, 1991, pp.2336-2339.
[5] E.Bonabeau, L.Dagorn, Possible universality in the size distribution of fish schools, *Phys. Rev. E.*, vol.51, 1995, R5220-5223.
[6]H.Takayasu, I.Nishikawa, H.Tasaki, Power-law mass distribution of aggregation systems with injection, *Phys. Rev. A*, vol.30, 1988, pp.3110-3117.
[7] H.Takayasu, Steady-state distribution of generalized aggregation system with injection, *Phys.*

Rev. Lett., vol.63, 1989, pp.2563-2565.

[8]S.K.Freedlander, *Smoke, Dust and Haze*, Willey-Int. Sci., New York, 1977.

[9] B.B.Mandelbrot, *The fractal geometry of nature*, W.H.Freeman, San Francisco, 1982.

[10] H.Takayasu, H.Inaoka, New type of self-organized criticality in a model of erosion, *Phys. Rev. Lett.*, vol.68, 1992, 966-969: R.Li, H.Takayasu, H.Inaoka, Water erosion on fractal surface, *Fractals*, vol.4, 1996, 385-392.

[11] H.Inaoka, H.Takayasu, Water erosion as a fractal growth process, *Phys. Rev. E*, vol.47, 1993, pp.899-910.

[12] J.J.Gilvarry, B.H.Bergstrom, Fracture of Brittle Solid.II.Distribution Function for Fragment Size in Single Fracture (Experimental), *J. of Applied Physics*, vol.32, 1961, pp.400-410.

[13] H.Inaoka and H.Takayasu, Universal fragment size distribution in a numerical model of impact fracture, *Physica A*, vol.229, 1996, 5-25.

[14] W.E.Leland, M.S.Taqqu, W.Willinger, D.V.Willson, On the self-similar nature of ethernet traffic (extended version), *IEEE/ACM Trans. Net.*, vol.2, 1994, 1-15.

[15] Misako Takayasu, Hideki Takayasu, Takamitsu Sato, Critical behaviors and 1/f noise in information Traffic, *Physica A*, to appear.

[16] Misako Takayasu, Characterization of violent fluctuations by interval distributions of level sets, *Physica A*, vol.197, 1993, pp.371-378.

Diversity and Complexity of Patterns in Nonequilibrium Systems
— Pattern Formation in Electrohydrodynamic Instability —

Shoichi Kai

Department of Applied Physics, Kyushu University, Fukuoka 812-81, Japan

Abstract

Pattern formation in nonequilibrium dissipative systems has attracted the interest of scientists for a long time, which is a very common phenomenon frequently observed in nature and shows rich diversity and complexity. The outline of their concept, classification and hierarchy is briefly discussed in general. As a concrete example, the electrohydrodynamic pattern formation in liquid crystals is introduced. It shows rich variety of phenomena and is easily accessible to various patterns by application of an ac voltage. Physics of the electrohydrodynamics, and their diversity and variety of pattern formation are described.

1. Introduction

Definition, quantization, characterization, formalization and classification of spatially complicated patterns are the areas to investigate in order to understand diversity and complexity of patterns as a science. The physical characterization of pattern complexity has made considerable progress during the recent two decades and is now one of the major branches of nonlinear science. This is largely due to the development of nonlinear nonequilibrium physics. Patterns in nonequilibrium dissipative systems appear via the balance between the dissipation and supply of energies and/or material mass. It is a branch of physics where in particular cases, for instance pattern formation in liquid crystals, a lively interaction between different disciplines, from applied mathematics to material science, exist. Fluid dynamics and turbulence, interface motion during solidification, chemical reaction and biological pattern formation have a similar impact on the conceptional development of the physics about systems far from thermal equilibrium. They all show quite interest diversity and complexity as we know.

For the closest way of one approach to understand a universality containing all these phenomena, better concept, better methodology and sophisticated mathematical tools should be found out. One candidate is "the so called reduction of information" which means that one only picks up spatially and temporally slow modes (envelope or amplitude) by elimination and renormalization of rapid modes (elements) after finding their hierarchy. The equations derived from this approach are often called the Ginzburg-Landau (GL) equation, the amplitude equation or the phase diffusion equation, of which dynamics are called the amplitude and phase dynamics. This approach is recently very successful to describe many macroscopic patterns and even spatiotemporally chaotic patterns in weakly nonlinear regimes.

One can meet pattern formation in various occasions, places and systems. For example, some cases are of temporal patterns such as in electric circuits, and other cases are of spatial patterns such as in snow crystals and biological morphogenesis. Physics for these formation problems is to understand individual mechanisms (individuality) of such objects. On the other hand, according to another classification, there may be of two cases; periodic and nonperiodic patterns (or ordered and disordered patterns). This classification contains rather universal problems independent of individuality of physical mechanisms. That is, it is closely related to the principle governing nature, *e.g.*, what is the origin of randomness or how order appears.

The main topics discussed in the present article is the pattern formation in the electrohydrodynamic convection (EHC) in nematic liquid crystals. It is a very convenient system to study pattern formation phenomena and shows rich variety of spatial and temporal patterns. As described above, furthermore, recent developments of research on patterns in liquid crystals are also closely related to other phenomena such as isotropic convective systems (*e.g.* Rayleigh-Bénard convection), crystal growth, interfacial motions between two phases, topological instabilities, phase transitions and so on. In theoretical and mathematical points of view, all the pattern formation must be similar and the similar approach may be available. Under such backgrounds of concepts, we will describe the variety and complexity of pattern formation in EHC in liquid crystals as an example.

2. Element, Wholeness and Hierarchy

In general, as described above, systems consist of elements. That is, physical systems are usually assembled of one or a few different varieties of elements with a large number of those elements. The property of a whole system is determined from the following factors of elements; (1)number of elements, (2)varieties of elements and (3)property of interactions among elements constructing the system. Only if the system consists of a large number of identical elements without interaction among them, it will be "summative" as a whole. Then one calls this property of the system "summativity". When the system consists of a large number of a few different elements which have some interactions among them, on the other hand, it is called the complex system. If it consists of many different varieties of elements in addition to a large number of elements, they are called the composite system. Both these systems are "constructive" because the whole is more than the sum of its parts (elements). Thus in many systems the wholes themselves become elements of a next order, that is, there is "hierarchy".

These constructive systems show the variety of behavior such as self-organization phenomena and instabilities. The most familiar and studied example related to an instability and self-organization phenomenon will be a phase transition.

Let us take here as an example a phase transition in magnetic materials in which the transition takes place between paramagnetic (disordered state) and ferromagnetic phase (ordered state); see Fig.1. Then a magnetic spin can be regarded as an element and the whole consists of many these elements (magnetic spins) with interactions. Due to thermal fluctuations at finite temperature T magnons are excited in such systems. For

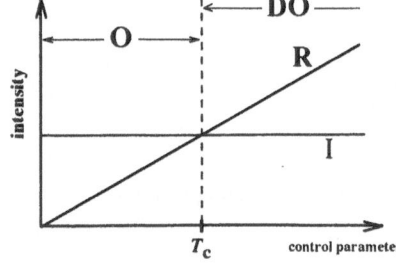

Fig. 1   Phase diagram due to competition between random force and interaction force. **R**: random force. **I**: interaction force. T_c: phase transition (instability) point. **O**: ordered state. **DO**: disordered state.

the higher temperature the more magnons are excited. Then the magnetic transitions take place collectively in the whole sample on a macroscopic scale as changing T. On this phenomenon the following consideration can be possible.

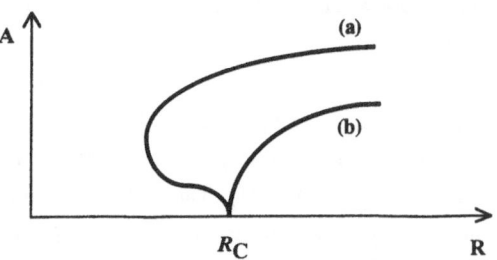

When a random force R due to thermal fluctuations is stronger than an interaction force I at higher temperature T than T_c, the system becomes disordered state (**DO**: paramagnetic phase), as shown in Fig.1. Here T_c is a phase transition temperature. When R is suppressed

Fig. 2   Bifurcation diagram in nonequilibrium state. (a) subcritical (b) supercritical bifurcation. R: control parameter (Rayleigh number in convective instability), A: order parameter.

by reduction of temperature, a ferromagnetic phase (**O**: ordered state) is realized (see. Fig.1). There are essentially two types of phase transitions (bifurcations); a first order (subcritical bifurcation, see Fig.2(a)) and a second order one (supercritical bifurcation; Fig2(b)). A transition between para- and ferromagnetic phases is usually a second order (continuous). This argument however only discusses the macroscopic property of the whole (wholeness).

Now we are going into conceptual but more detailed argument how each elements organize to form the ferromagnetic phase looking at microscopic levels. Now let A_s, $\chi_s \propto 1/T$, f and β be the response, a magnetic susceptibility for a single element, external forces and the interaction strength

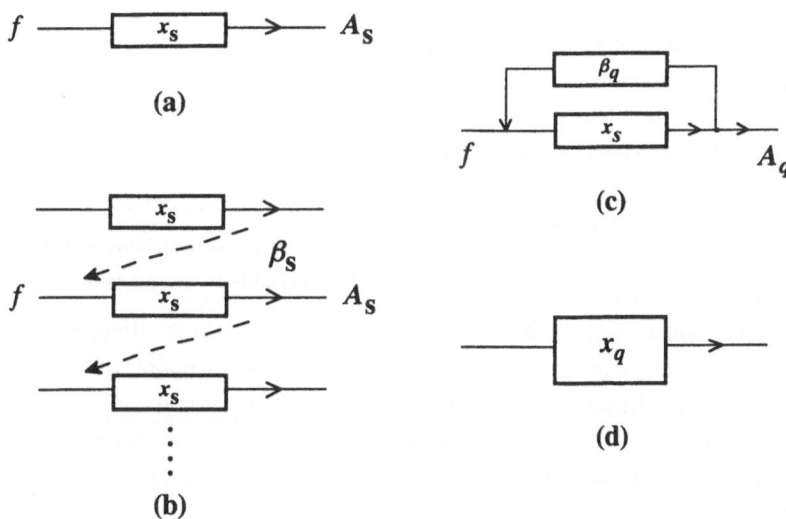

Fig. 3   Renormalization in system with many elements. (a) single elements approach. (b) mean field approach by a feed-back process. (c) renormalization by collective modes q. f: random force, χ_s: susceptibility for a single element, χ_q: susceptibility for collective modes, A_s: response of a single element, A_q: response of collective modes, β_q: interaction of collective modes.

respectively, as shown in Fig.3a. Then at $\beta = 0$ the relation for a large number of such elements (Fig.3b),

$$A_s = \chi_s f \tag{1}$$

can hold. There is no instability in this relation because of no singularity in χ_s except $T = 0$. When there is no interaction among elements, each elements behave independently without any collective motions according to Eq.(1). That is, the entire response will be vanished because of simple summation of random responses (summative), *i.e.* a paramagnetic phase (disordered state, **DO** in Fig.1).

In the presence of interactions among elements or parts, *i.e.* $\beta_q \neq 0$, and supposing that each elements equally feel their self-inducing fields (the so-called mean field approximation; see in Ref. [3]), Fig.3b can be rewritten as Fig.3c and the relation can be also given by the equation;

$$A_q = \frac{\chi_s}{1 - \chi_s \beta_q} f = \chi_q f. \tag{2}$$

Here A_q represents the collective (entire) response and a subscript q indicates variables for q-collective modes as well as s for a single element. Now the susceptibility of the whole is given by renormalization as $\chi_q = \chi_s / (1 - \chi_s \beta_q)$ (see Fig.3d). Figure 3d shows such a renormalization process where one obtains again similar description in Fig.3a (*i.e.* a kind of hierarchy). (Thus an oscillatory instability in electric circuits has an analogy with phase transitions as schematically shown in Fig.4 (as an analogy for Fig.3c). That is, the oscillatory instability is a kind of phase transitions.)

The instability (phase transition) takes place at $\beta_q = 1 / \chi_s (T = T_c)$. Because of $\chi_q = \infty$, then, an infinitesimal external force (signals) leads to $A_q = \infty$, that is, new macroscopic order starts. Then A_q becomes the order parameter A

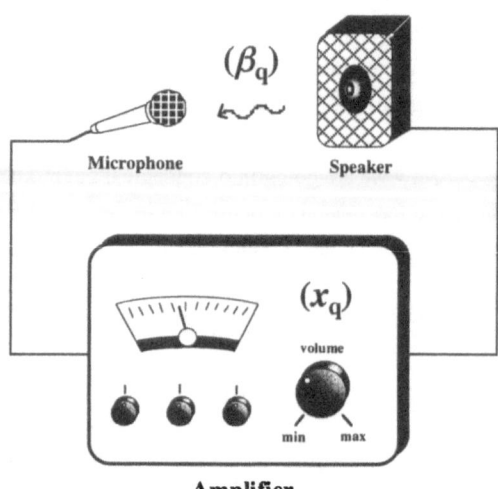

Fig. 4   Oscillatory instability in an electric amplifier. β_q: feedback rate, χ_q: amplification rate.

of which linear response may be written as $\partial A / \partial t = \chi_q^{-1} A$ since $f = \partial A / \partial t$. However the response of the ordered state is not linear anymore in general and nonlinear description will be necessary, described by *e.g.* the time-dependent GL equation. If one takes further macroscopic scales such as a set of a large number of A, each A may become the element for the whole in the next scale. Such phenomena and hierarchy are frequently observed in systems far from equilibrium, for example in convective instabilities.

3. Spatiotemporally Regular and Chaotic Patterns

In the above section, our description has been concentrated to periodic ordered structures. There are, however, nonperiodic and disordered patterns which are rather more common in nature. In order to describe such varieties of patterns, the so-called complex Ginzburg-Landau (CGL) equation can be sometimes used;

$$\frac{\partial A}{\partial t} = (1 + ic_0)A + (1 + ic_1)\nabla^2 A - (1 + ic_2)|A|^2 A. \tag{3}$$

The CGL is originated from the complex amplitude equations which describe the slow modulation in space and time near the threshold for an instability, and describes the chaotic behavior as well as fronts, pulses, sources, sinks and waves. Small disturbances can grow for $1 + c_1 c_2 < 0$ in CGL, which is called the Benjamine-Feir instability, a kind of phase instabilities (corresponding to the Eckhaus instability in a real GL equation), and frequently leads to chaotic states (spatiotemporal chaos: STC). The phase diagram is schematically drawn in Fig.5. There are several types of STCs, such as amplitude, phase and defect chaos (defect turbulence; see Fig.6), where amplitude, phase ,and both amplitude and phase become random in space and time, respectively. Here defects are characterized by zero amplitude and 2π-phase shift at their cores. We exclude fully developed turbulence and only include turbulence close to

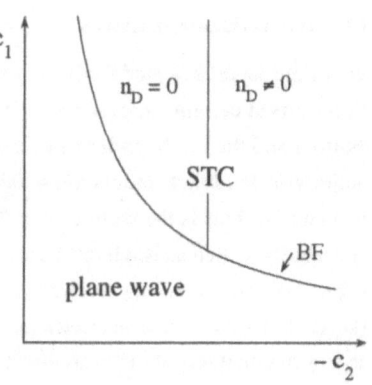

Fig. 5   Rough schematic drawing of phase diagram for one-dimensional complex GL equation in c_1 - c_2 plane. **BF**: the Benjamin-Feir instability limit. n_D: number of defects.

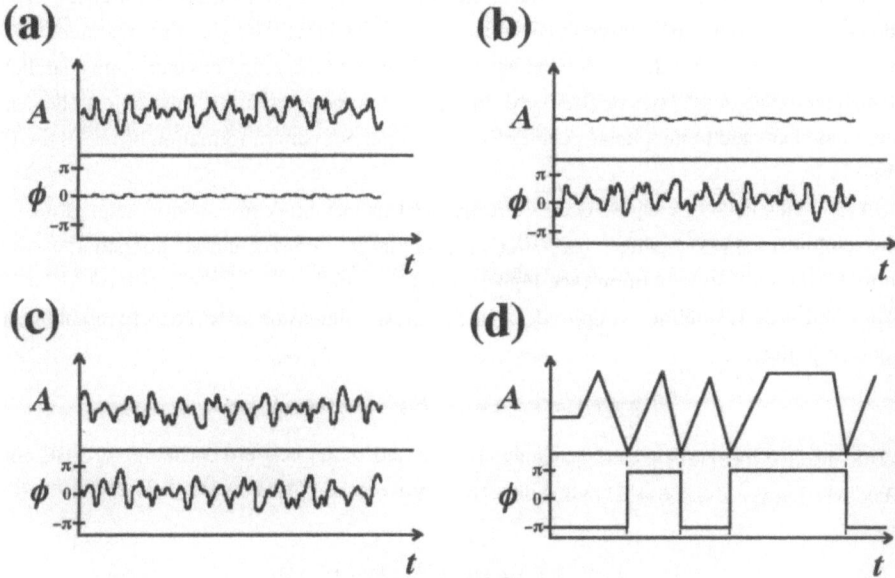

Fig. 6   Schematic drawing of amplitude (a), phase (b), both amplitude and phase (c), and defect turbulence (d).

an instability point (weakly nonlinear regime). Typical examples of these patterns are observed and well-studied in convective systems where the element for the whole is an each convection roll. One of them is the electrohydrodynamic convection (EHC) in liquid crystals which will be described in the following section.

4. Electrohydrodynamics

When a fluid layer is heated from the below, fluid convection starts at certain temperature difference between the bottom and the top boundaries of a layer, due to an autocatalytic feedback mechanism described above (Figs.3 and 4). This is the well-known thermal convective instability called as Rayleigh-Bénard (RB) convection (Fig.7a). Similarly, the electrohydrodynamic convection (EHC) occurs when an electric ac-voltage V stronger than a certain threshold V_c is applied to a thin nematic layer with a negative dielectric constant $\varepsilon_a = \varepsilon_{//} - \varepsilon_{\perp} < 0$ [6]. Here $\varepsilon_{//}$ and $\varepsilon_{\perp}$ are the dielectric constants parallel and perpendicular to the director, respectively, which rep-

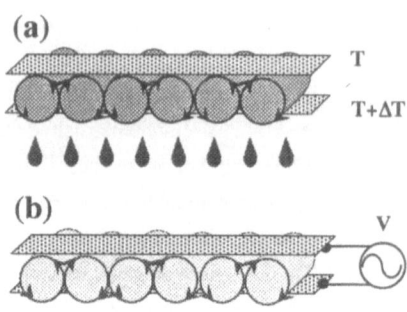

Fig. 7   Convections in Rayleigh-Bénard instability (a) and EHC (b).

resents the averaged direction of the molecular orientations of liquid crystals [6]. These are typical self-organization phenomena and one of the most convenient systems to investigate pattern formation in systems far from equilibrium [1, 6, 9]. Quite interesting hierarchy has been also reported [7].

The most advantageous fact of EHC is the easy control of their initial symmetries. That is, liquid crystal molecules can be forced to be aligned to desired directions by simple treatments of glass substrates, for example the planar and homeotropic alignments where the director aligns parallel and perpendicular to the glass plates respectively. In many past investigations on this subject, then, much attention has been paid to the planar geometry and its variety of pattern formation has been found out [1, 9].

EHC in the planar orientation occurs directly as a first instability from non-structured state. The weakly nonlinear aspect of pattern formation near the onset of convection is understood by experimentally taking the Busse balloon (see Fig.8). The Busse balloon is also mapped out by stability analysis of the weakly nonlinear amplitude equations in two dimension after properly rescaling space, time and amplitude;

$$\frac{\partial A}{\partial t} = \varepsilon A - |A|^2 A + (\partial_x^2 - ik_2 \partial_x \partial_y^2 + w\partial_y^2 - \partial_y^4)\, A, \tag{4}$$

and the detailed comparison can be available. Here k_2 and w are non-zero constants for EHC and for RB convection $k_2 = 2$ and $w = 0$. In one dimensional systems, Eq.(4) shows no difference for EHC and RB;

$$\frac{\partial A}{\partial t} = \varepsilon A - |A|^2 A + \partial_x^2 A. \tag{5}$$

Here A is the envelope (amplitude) function of, *e.g.*, the convective velocity u; $u = \varepsilon^{1/2}[A \exp(iq_c x) +$ c.c.] $+ \, \mathrm{O}(\varepsilon)$. From the linear analysis of Eq.(5) the neutral stability (NS) and the Eckhaus (E) bound-

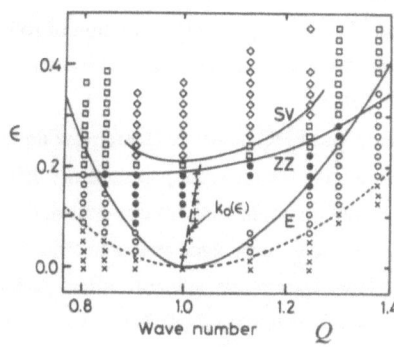

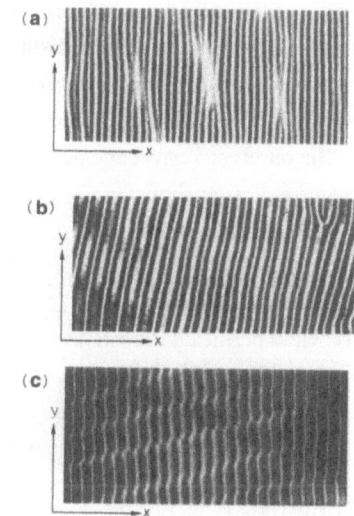

Fig. 8   Busse balloon in planar EHC. **E**: the Eckhaus instability((a)) **ZZ**: zig-zag instability((b)) **SV**: skewed varicose instability((c)) The dotted line is a marginal stability limit.

aries are theoretically determined [2, 10]. Here NS gives dispersion relation of the short wavelength roll convection, and E is a long wavelength and a secondary instability.

An experimental example is shown in Fig.8 where all weakly nonlinear aspects are displayed in two dimensional EHC systems. Here Q and ε indicate the values of deviation from the critical wavenumber q_c, and of deviation from the convective onset as $\varepsilon = (V^2 - V_c^2) / V_c^2$, respectively. Examples of images of these instabilities are shown in Fig.8 simultaneously. In a two dimensional planar system,

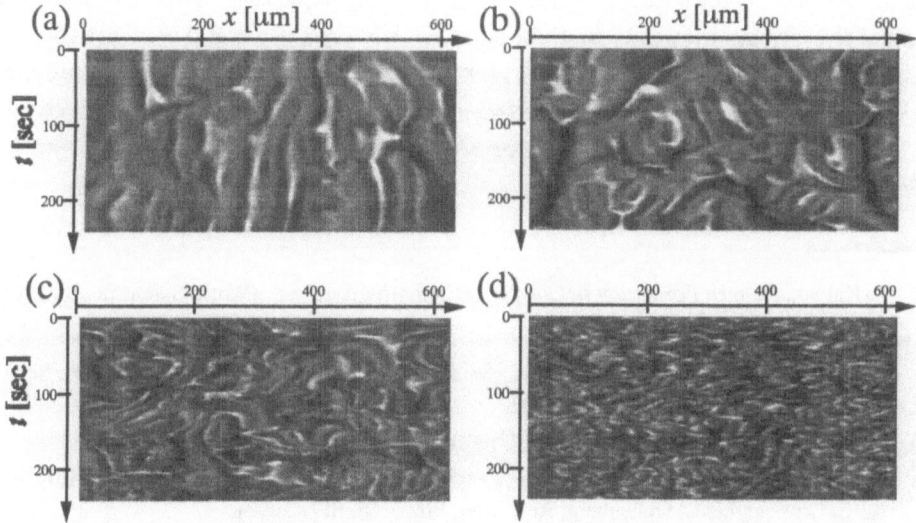

Fig. 9   Spatiotemporal map of weak turbulence in homeotropic EHC. (a) $\varepsilon = 0.1$ (b) 0.19 (c) 0.33 (d) 0.81.

one observes basically three different selection dynamics for EHC pattern formation such as (1) simple decay of rolls, (2) elastic relaxation of rolls (in some cases by defects) and (3) relaxation of rolls via additional instabilities, such as zig-zag, Eckhaus and skewed-varicose instabilities. The corresponding transition lines define the Busse balloon in the planar EHC.

The observed convective pattern at onset in homeotropic EHC however is irregular and non-periodic, i.e. STC as shown in Fig.9 (see Ref [10, 11] for detail). The tendency is certainly different from STC in the planar case, where the onset of STC does not coincide with the convective onset as well as conventional STC of complex TDGL equations. A route to turbulence and chaos via a single supercritical bifurcation has been observed so far in very few experiments, whereas direct transition to STC via subcritical bifurcations from non-convective (or non-structural) state have been observed more often. The Busse balloon in a two dimensional homeotropic system however cannot be obtained, i.e. there is no Busse balloon. Thus the symmetry of systems plays an important role for pattern formation and frequently leads to qualitatively different phenomena.

5. Conclusion

Diversity and complexity are very closely related to nonlinearity governing phenomena. Nonlinearity of phenomena leads to individuality as well as universality. Individuality in some sense corresponds to diversity. So far physics has traditionally developed by reconstruction of objects after once dividing each elements and by formulating of basic governing equations for them. In particular the under-laid universality has been sought and in many case it was successful, for example in critical phenomena. However, nonlinearity usually does not allow to treat it by the conventional methodology. We have known that from discovery of chaos and its property the common knowledge in traditional physics cannot be totally acceptable because deterministic equations show nonperiodic features ,and that nonlinearity often leads to nontrivial and surprising phenomena. Thus nonlinearity is the origin of diversity, hierarchy and complexity in nature. It is a first step to study pattern formation in weakly nonlinear regimes as described in the present article in order to understand full nonlinearity in the future. The pattern formation even in weakly nonlinear regimes however shows quite rich varieties and complexity, as is understood from concrete examples in EHC. I hope that most readers are interested of the subjects.

References

[1]   S. Kai ed., *Pattern Formation in Complex Dissipative Systems*, (World Scientific, Singapore, 1991).

[2]   P. E. Cladis and P. Palffy-Muhoray ed., *Spatio-Temporal Patterns*, (SFI Studies in the Sciences of Complexity, Addison-Wesley, 1995).

[3]   H. Thomas, IEEE Trans. Mag., **5**, 874 (1969).

[4]   P. Manneville, *Dissipative Structures and Weak Turbulence* (Academic Press New York 1990).

[5]   M. C. Cross and P. C. Hohenberg, Rev. Mod. Phys., **65**, 851 (1993).

[6]   P. G. de Gennes, *The Physics of Liquid Crystals* (Clerendon Oxford 1974).

[7]   S. Kai and K. Hirakawa, Prog. Theor. Phys., supp. **68**, 212 (1978); S. Kai and W. Zimmermann,

Prog. Theor. Phys. supp. **99**, 458 (1989).

[8] L. Kramer, E. Bodenshatz, W. Pesch, W. Thom and W. Zimmermann, Liq. Cryst., **5**, 699 (1989).

[9] L. Kramer and W. Pesch, Annu. Rev. Fluid Mech., **27**, 515 (1995).

[10] S. Kai, K. Hayashi and Y. Hidaka, J. Phys. Chem., **100**, 19007 (1996).

[11] See in this book, Y. Hidaka *et al.*

Complex Motion of Twin Boundary
in Molecular Crystal (TMTSF)$_2$PF$_6$

MUKOUJIMA Mika, KAWABATA Kazushige and SAMBONGI Takashi

Department of Physics, Graduate School of Science, Hokkaido University,
Sapporo 060, JAPAN

Abstract

In molecular crystals (TMTSF)$_2$PF$_6$, a twin boundary shows complex motion under controlled external force. The twin boundary moves intermittently and its velocity varies from run to run though same measurement was repeated under the same force and within the same area of the crystal. When position dependent velocity of the twin boundary is normalized by the average velocity in each run, the normalized curves coincide each other. This result means that the complex motion is subject to two characteristic factors; one determines the pattern of motion and the other does the magnitude of average velocity.

Key words: twin, kink, plastic deformation, molecular crystal, organic crystal

1. Introduction

Plastic deformation of crystals occurs by two mechanisms; motion of dislocation and twinning. The deformation is developed through collective simultaneous motion of a macroscopic number of molecules or atoms on the boundary. The motion of an isolated twin boundary has not been investigated systematically, while dynamics of dislocation has been examined in detail for some decades.

Twin-deformation in the molecular crystal of (TMTSF)$_2$X, where TMTSF denotes tetramethyltetraselenafluvalene and X=PF$_6$, AsF$_6$, ClO$_4$ etc., was first reported by Schwenk *et al.* [1]. They found that twin-deformation occurs easily and twin boundary (referred to as "kink" hereafter) can be moved along the needle axis under small lateral force. Ishiguro *et al.* [2] found from analysis of X-ray precession photographic images, that two sides separated by a kink are in mirror symmetry with respect to the kink plane.

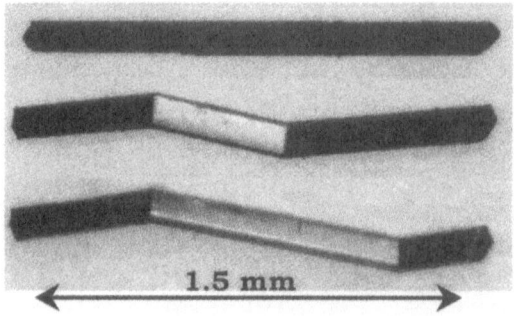

Fig.1: <u>Top</u>: A single crystal of (TMTSF)$_2$PF$_6$.   <u>Middle</u>: A kink pair formed by external force.
<u>Bottom</u>: The right kink is moved by applying external force.

In our previous works, dynamics of the isolated kink induced in single crystals of (TMTSF)$_2$PF$_6$ under lateral external force was studied [3,4]. We found that the kink moves intermittently under constant stress; the kink moves and stops, and the kink motion is resumed after some intermission. The kink velocity varies from run to run though repeated measurements was carried out under the same force and within the same area of a sample. The pattern of kink velocity, however, closely resembles each other. In this paper we report the repeated kink motion without intermission.

2. Experimentals

Single crystals of (TMTSF)$_2$PF$_6$ are needle-shaped triclinic ones and the needle axis is parallel to the [100] direction. When sharing force is applied perpendicular to the needle axis, twin-deformation is formed in a part of crystal and two corresponding kinks appear parallel to (210). Figure 1 illustrates that twin deformation occurs in the middle part of a crystal and the right kink is moved by applied external force. The position of a kink was measured by means of high-speed CCD digital camera and an optical microscope at room temperature. External force was kept constant within 1% during kink motion. After a run, the moved kink was pushed back to the initial position, and then the measurement was repeated in the same area of the crystal. Details of our experimental technique have been published previously [3].

3. Results and Discussion

Time dependent position of the kink under constant stress (7.9×10^4 Pa) is shown in Fig. 2. Two characteristics are apparent in the figure. The first is that the kink moves intermittently; the kink moves and stops, and its motion resumes after intermission. The positions where intermissions occur (near the position of x=0.13mm and x=0.21mm in Figure 2) are reproduced in every run. So the kink must be pinned strongly by barriers such as localized defects. The kink which is stopped by barriers can not pass through the pinning sites as long as external stress is constant, if it is a planer rigid object. The second characteristic is that the kink position vs time curve varies from run to run even though measurements was carried out under the same force and within the same area.

Hereafter, we examine the second characteristic in detail. We measured the kink motion with higher time resolution in an area where no intermission was observed. Figure 3 shows the typical time-dependent position under the constant stress of 8.4×10^4Pa. Figure 4(a) shows the kink velocity calculated from these curves as a function of its position, so as to compare different runs without intermission.

Although the magnitude of the local velocity varies from run to run, the pattern of kink velocity closely resembles each other; a kink moves more slowly in some area while more rapidly

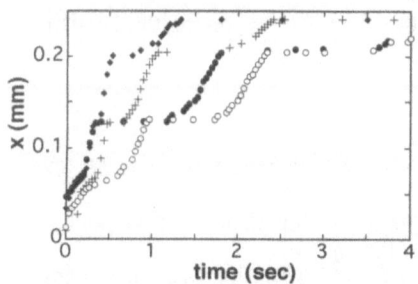

Fig.2: Time-dependent position of a kink
under constant external stress

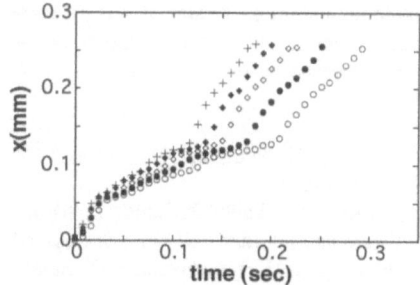

Fig.3: Time-dependent position of a kink
between intermissions

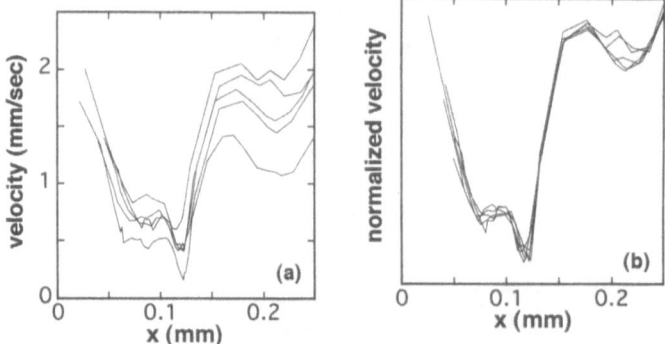

Fig. 4: (a) Position dependence of kink velocity for five runs
(b) Kink velocity normalized by each average velocity

in other area. Since relatively high (low) velocity is always reproduced in the definite area, the local velocity vs position depends on imperfection inside the crystal such as defects or impurity invading during crystal growth. Figure 4(b) shows the position dependence of kink velocity, which is normalized by the average velocity $\overline{V}$ of each run. It is found that the normalized curves coincide each other. This result means that the complex kink motion is subject to two factors. The first one determines the pattern of motion and the other determines $\overline{V}$ for each run. The latter factor keeps persistent influence on the kink motion through the run. While $\overline{V}$ varies even under the constant stress, the average of $\overline{V}$ among a number of runs increases monotonously with stress. If the kink velocity is determined by external stress uniquely, the force should fluctuate by as much as 20% from run to run. Thus fluctuation of $\overline{V}$ is not caused by uncontrolled fluctuation of applied stress. Although the origin for such a factor is not clear at present, we consider that it is related to internal collective phenomena such as rearrangement of molecules on the twin boundary or redistribution of imperfections scattered on some area. The average kink velocity in each run does not change in a systematic way during repeated runs. Thus the variation of $\overline{V}$ is not caused by mechanisms such as gradual breaking of a crystal, or gathering of imperfections inside some area.

The measurement was carried out under different stress between 7.7×10^4 and 9.0×10^4 Pa and in other crystals. We found that the velocity vs position curves resemble each other even under different stress. In addition, the normalized curves under different stress coincide each other with suitable $\overline{V}$. Thus the factor which determines $\overline{V}$ is universal and independent of applied stress.

In conclusion, we found that the motion of twin boundary in $(TMTSF)_2PF_6$ is intermittent and the average velocity $\overline{V}$ between intermissions varies from run to run even when the measurements was carried out under the same condition. However, the kink velocity in each run can be normalized by $\overline{V}$ and the normalized curves coincide each other.

Acknowledgment- The author (K.K) is grateful for the financial support of the Ministry of Education, Science and Culture of Japan (No. 08455048).

References

[1] H. Schwenk, K. Neumaier, K. Andres, F. Wudl and E. Aharon-Shalom, *Mol. Cryst. Liq. Cryst.* **79**, 277 (1982)
[2] T. Ishiguro, T. Ukachi, K. Kato, K. Murata, K. Kajimura, M.Tokumoto, H. Tokumoto, H.Anzai and G. Saito, *J. Phys. Soc. Jpn.* **52**, 1585 (1983)
[3] M. Mukoujima, K. Kawabata and T. Sambongi, *Solid State Commun.* **98** No.4, 283 (1996)
[4] M. Mukoujima, K. Kawabata and T. Sambongi, *J. de Phys. I* **6**, (1996) in press.

Dynamic Scaling of etched surface roughness of LaSrGaO$_4$ single crystal by Atomic Force Microscopy

Yuzuru Takagi[1], Satoshi Tanda[1], Noriyuki Fujiyama[1], and Ikuto Aoyama[2]

1 Department of Applied Physics, Hokkaido University, Sapporo, 060, Japan
2 Research Division, Komatsu Ltd., 1200 Manda, Hiratsuka-shi, Kanagawa 254, Japan

Abstract

We have studied kinetic roughening of etched surfaces of LaSrGaO$_4$ single-crystals, having layered perovskite structure like a high-T$_c$ cuprate, by using atomic force microscopy. We found that the surface roughness exhibited a power law behavior on both the length scale and the etching time, which have self-affine structures such as surfaces of growing films. From our experiments, the etched LaSrGaO$_4$ surface can be described as a self-affine fractal with a roughness exponent $\alpha = 0.90 \pm 0.06$ and a growth exponent $\beta = 0.33 \pm 0.05$. Our results provide the first evidence that the etching process of LaSrGaO$_4$ surface is dominated by a *lattice vacancy* growth model of the molecular beam epitaxy(MBE) type.

Key words: dynamic scaling, roughness, surface, self-affine, atomic force microscopy

1.Introduction

The nonequilibrium dynamics of surface growth has attracted much attention in recent years. Especially, growth processes such as vapor deposition, molecular beam epitaxy(MBE) and electrodeposition have been investigated by using dynamic scaling theory[1][2]. Theoretically, Kardar-Parisi-Zhang(KPZ) studied growing interface based on the nonlinear Langevin equation using dynamic renormalizaton group[3]. The ballistic deposition model and Eden model are believed to belong to the same universality class as KPZ growth[4]. Many theoretical and numerical simulation studies have been investigated to obtain a more complete picture of possible universality classes of growing surfaces. However, not much attention was paid to the research of surface roughness of etched processes by using the dynamic scaling theory so far, therefore neither experimental nor theoretical data have been found in the literature concerning the application of dynamic scaling to etching surfaces. In principle, we think that the analysis of the etched surfaces can be made using the dynamic scaling theory. In this paper, we report the roughness evolution during etching LaSrGaO$_4$ single-crystal surfaces under nonequilibrium conditions, followed by atomic force microscopy(AFM) images. Furthermore we investigated etched surface roughness of Si and compared etched surface roughness of LaSrGaO$_4$ to etched surface roughness of Si. From the observed results, it found that there was a difference between isotropic etching processes and anistropic etching ones. We suggest that the etching process of LaSrGaO$_4$ surface is described as a *lattice vacancy* growth model of the MBE type.

2.Experimental and Discussion

There are many chemical etching procedures, but we do not compare procedures in this paper. Rather, we used the easy procedure below because we found it repeatable and controllable. LaSrGaO$_4$ single-crystal, which has layered perovskite structure, was etched in a stirred 0.1 % HNO$_3$ and CH$_3$OH solution at T = 298 K. We made samples for etching times of 0, 5, 10,

50, 100, 500 min. To compare layered materials with isotropic ones, the chemical etching was carried out on (100) oriented Si single-crystal (4.5-8.5 Ωcm, p-type) wafers at T = 292 K. Si wafers were immersed on an aqueous solution containing 0.1M NaOH. We made samples for etching times of 1, 2, 3, 5, 8, 10 min. We measured etched surface shapes by AFM. AFM is not only capable of high 3D resolution observation of surfaces, but also capable of observation of wider areas, and is more useful than scanning electron microscopy(SEM) and transmission electron microscopy(TEM). We performed measurements on a variety of area sizes, i.e., $0.5 \times 0.5, 2 \times 2, 5 \times 5 \mu m^2$. In each measurement, the height data were acquired at a resolution of 256×256 pixels. Figure 1 shows the AFM perspective view images of etched surfaces of LaSrGaO$_4$ for 10, 50, 100 min. As the etching time increases, the surface becomes rougher.

(a)   (b)

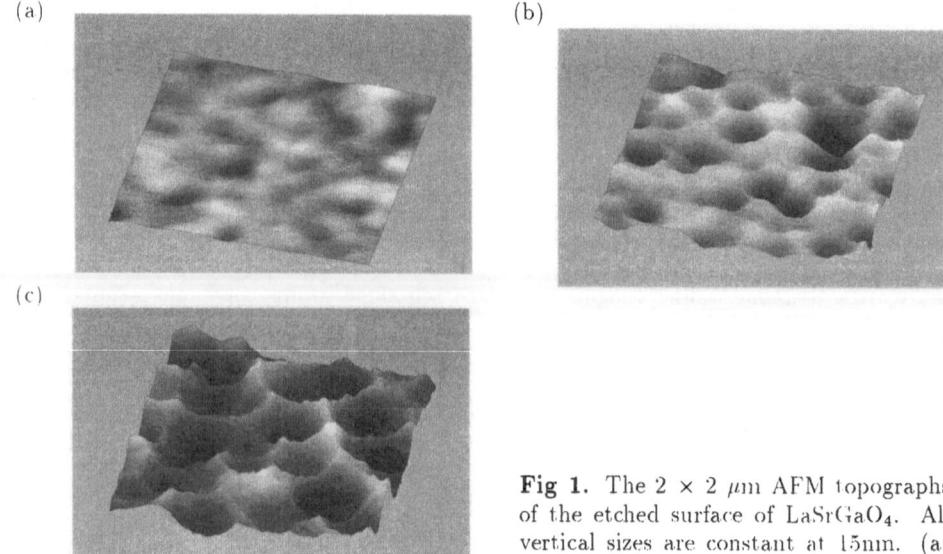

(c)

Fig 1. The 2×2 μm AFM topographs of the etched surface of LaSrGaO$_4$. All vertical sizes are constant at 15nm. (a) 10min (b) 50min (c) 100min.

In order to estimate of critical exponents of etched surface roughness, we analyze with finite size scaling of growing processes of films by deposition. This scaling theory predicts that the surface width W scales with etching time t and length scale L as

$$W(L, t) \propto L^\alpha f(x). \tag{1}$$

where W(L,t) is defined by

$$W(L, t) = (\frac{1}{L^2}\Sigma[h(x_i) - \langle h \rangle]^2)^{\frac{1}{2}}. \tag{2}$$

$h(x_i)$ is the deposit height at site i, $\langle h \rangle$ is the mean height and $x = t / L^{\frac{\alpha}{\beta}}$. where α, β is roughness and growth exponents, respectively. Furthermore, $f(x)$ has the following properties: $f(x)$ = const for $x \Rightarrow \infty$ and $f(x) = x^\beta$ for $x \Rightarrow 0$. Equation (1) consists of two limiting situations, namely, when $t \to 0$, Eq.(1) is reduced to

$$W(L, t) \propto t^\beta. \tag{3}$$

On the other hand, for $t \to \infty$, Eq.(1) is reduced to

$$W(L, t) \propto L^{\alpha}. \tag{4}$$

Figure 2(a),(b) shows the surface width $W(L, t)$ of etched LaSrGaO$_4$ and etched Si for each etching time. In case of LaSrGaO$_4$, we noted the 50 min or more etching samples because the value of α needs almost constant value to analyze with (1) scaling formula. For each etching time, the curve consists of two regimes separated by the correlation length L_c. fitting the low length scale behavior and the saturation behavior to straight lines. For the lower region, W scaled with length scale L as L^{α}. The roughness exponents α became constant at 0.90 ± 0.06 and 0.84 ± 0.04, respectively. Figure 3(a)(b) shows the saturated value of the surface width W_{sat}. In the upper region, W_{sat} scaled with the etching time t as t^{β}. The growth exponents of LaSrGaO$_4$ and Si became constant at 0.33 ± 0.05 and 0.64 ± 0.03, respectively. From the measured exponents, we found that there was a difference between the critical exponents of LaSrGaO$_4$ and those of Si. It means that etching processes of anisotropic materials such as LaSrGaO$_4$ belong to a universality class which is distinct from that of isotropic materials.

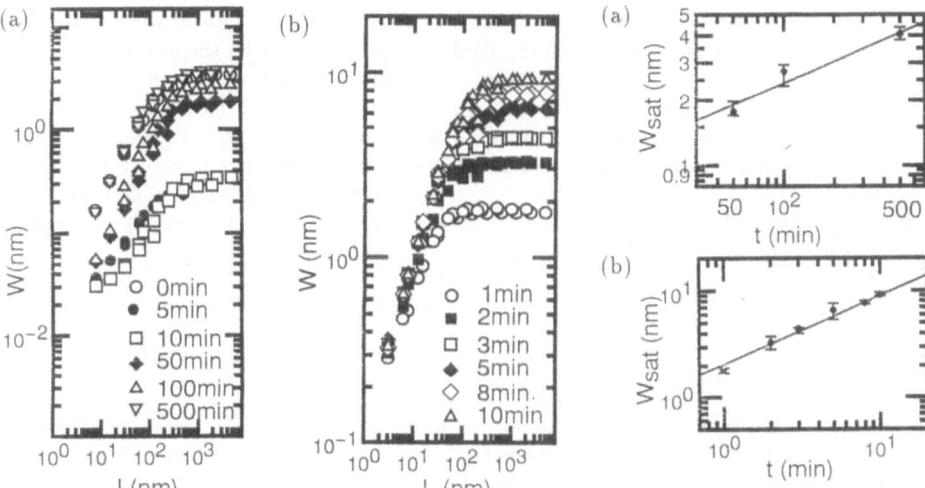

Fig2(a),(b). The surface width W vs length scale L. (a)LaSrGaO$_4$ single-crystals (b)Si single-crystals

Fig3(a),(b). The saturated width W$_{sat}$ vs etching time t.(a)LaSrGaO$_4$ single-crystals (b)Si single-crystals

The critical values of LaSrGaO$_4$ are in good agreement with the MBE growth model[5]. Accordingly, we suggest that the etched surface roughness of LaSrGaO$_4$ is dominated by the lattice vacancy growth of MBE process. Further studies are needed to see whether the etched LaSrGaO$_4$ is described by the lattice vacancy type of MBE growth. We thank Prof.Yamaya for useful discussion. We also thank T.Teramoto for experimental support.

References

[1]R.C.Salvarezza, L.Vázquez, P.Herrasti. P.Ocón, J.M.Vara, and A.J.Arvia
   Europhysics Letters, Vol.15, pp.727(1992).
[2]A.Iwamoto, T.Yoshinobu, and H.Iwasaki, *Physical Review Letters*, Vol.72, pp.4025(1994).
[3]M.Kardar, G.Parisi, and Y.C.Zhang, *Physical Review Letters*, Vol.56. pp.889(1986).
[4]T.Vicsek. *Fractal Growth Phenomena*, World Scientific, Singapore. 1990.
[5]D.E.Wolf, and J.Villian, *Europhysics Letters*, Vol.13, pp.389(1990).

Scaling behavior of the surface roughness in the crystallization process :transformation of polycrystalline Si from amorphous Si

N. Fujiyama[1], S. Tanda[1], Y. Takagi[1], K. Sajiki[2], and Y. Niwatsukino[2]

1.Dept. of Applied Phys., Hokkaido Univ., Sapporo-shi, Hokkaido 060, JAPAN
2.Research Division, Komatsu Ltd., 1200 Manda, Hiratsuka-shi, Kanagawa 254, JAPAN

Abstract

We made polycrystalline silicon (poly-Si) thin films from amorphous silicon (a-Si) thin films by the excimer-laser crystallization method. And we have investigated the surface roughness in the process from a-Si to poly-Si crystallized gradually by changing the energy density E of the excimer-laser. By using atomic force microscopy (AFM), the surface width $W(L, E)$ is estimated as the root mean square (rms) value of the roughness, where L is the length scale, E the energy density of the excimer-laser. We discovered a new scaling law in the crystallization process. We observed $W(L, E) \sim L^\alpha$ ($L \ll L_x$), and the saturation width $W_{sat}(E)$ scaled with E as $W_{sat}(E) \sim E^{\beta^*}$ ($L \gg L_x$), where L_x is the crossover length, the roughness exponent $\alpha = 0.88 \pm 0.10$, and the exponent $\beta^* = 1.37 \pm 0.10$, respectively.

Key words : roughness, surface, crystallization, dynamic scaling, atomic force microscopy

1. Introduction

There are many problems in phenomena on surface and interface. For example, they are dynamical properties in the crystallization or the growth process. On the growth process, we know the morphology of the suface roughness generally depends on the length scale of observation. Recently, the scaling approach, which is applied to experiments and theories on equilibrium phase transitions, has become a standard tool in the study of growing surfaces. In particular, the dynamic scaling approach [1, 2] has been applied to the study of variety of experiments [2] and theoretical models [1] of growing surfaces. And in theoretical studies [2], many equations have been advocated, for example, the Karder-Parisi-Zhang (KPZ) equation [3], the Edwards-Wilkinson (EW) equation [4], and so on. Generally, the morphology is characterized by the surface width $W(L, t)$ defined as the root mean square (rms) value of the roughness, i.e., $W(L, t) = [1/L^2 \sum(h_i - h(t))^2]^{1/2}$, where L is the length scale, t is the evolution time, h_i is the height at site i, and $h(t)$ is the mean height within the square of size $L \times L$. $W(L, t)$ shows a power law behavior,

$$W(L, t) \sim \begin{cases} L^\alpha & (\ L \ll L_x \ ) \\ t^\beta & (\ L \gg L_x \ ), \end{cases} \tag{1}$$

where α is the roughness exponent, and β is the growth exponent. Having an accurate numerical values for the exponent is important for us in order to identify the universality class in the system. However, there doesn't exist a systematic formalism for treating the growth process of the surface. On the crystallization process, we have a little idea on the dynamic properties, for example, when nucleation take place, what shape the surface forms, and so on. We thought the dynamic scaling approach could apply to the study of crystallization, too. And we suggest the dynamic scaling approach is a new method for clarifying the crystallization process. The purpose of this study is determining the universality class for the crystallization process from amorphous silicon (a-Si) to polycrystalline silicon (poly-Si). We made the poly-Si thin films from a-Si thin films by the excimer-laser crystallization method. And we observed the surface roughness of poly-Si thin films using atomic force microscopy (AFM), and analyzed the surface roughness by the dynamic scaling.

2. Experimental

A-Si thin films were grown on a glass by the method of plasma chemical vapor deposition. The thickness was about 1000 Å, determined by the scanning electron microscope. We could find a-Si thin films are crystallized gradually, according to laser power. We made poly-Si thin films from a-Si thin films by irradiating the excimer-laser. The excimer-laser has a pulse duration of 15 ± 0.8 ns in full width at half-maximum. The wavelength is 248 nm. We made five samples for different energy density of the excimer-laser. The energy density of laser was 50, 100, 150, 200, 250 mJ/cm^2. The number of shot was one to each samples. We measured the surface roughness gradually changing the length scale using AFM. Figure 1 (a)-(c) shows 3D AFM images for 50, 150, 250 mJ/cm^2, respectively.

(a)

(b)

(c)

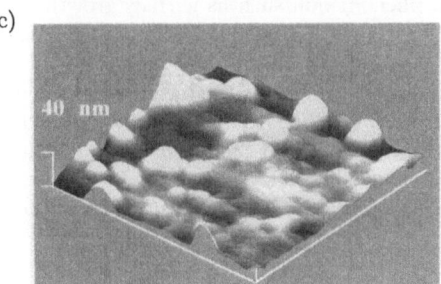

Fig.1. 3D AFM images for poly-Si thin films. All image sizes are $1 \times 1 \mu m^2$, and the energy density and actual vertical-axis scales are (a) $E = 50$ mJ/cm^2, 0-8 nm, (b) $E = 150$ mJ/cm^2. 0-20 nm and (c) $E = 250$ mJ/cm^2, 0-40 nm, r spectively.

3. Results and Discussions

The surface width $W(L, E)$ is defined as the rms value of the roughness calculated within a square of size $L \times L$, i.e., $W(L, E) = [1/L^2 \sum (h_i - h(E))^2]^{1/2}$, where E is the energy density of the excimer-laser, h_i is the height at site i, and $h(E)$ is the mean height within a square of size $L \times L$. Figure 2 shows the variation of the surface width $W(L, E)$ with the length scale L for five different energy density E, 50, 100, 150, 200, 250 mJ/cm^2. In spite of the difference in E, they exhibit a power low behavior. Each curve is separated into two regions by the crossover length L_x, fitting the low length scale behavior and the saturation behavior to straight lines. In the lower region, $W(L, E)$ shows a power law,

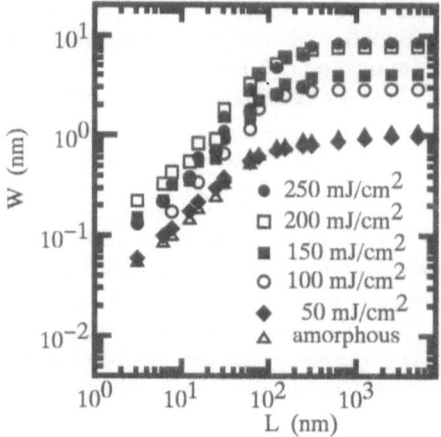

Fig.2. The surface width $W(L, E)$ of the poly-Si versus the length scale L plots calculated for all the samples.

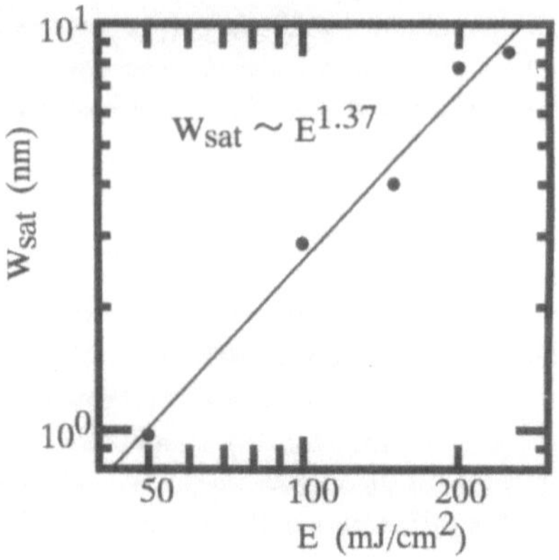

Fig.3. The saturation width $W_{sat}(E)$ versus the energy density E of the excimer-laser plots calculated for all samples.

$$W(L, E) \sim L^{\alpha} \quad (L \ll L_x), \tag{2}$$

where $\alpha = 0.88 \pm 0.10$. In the upper region, the saturation width $W_{sat}(E)$ shows a power law behavior (Figure 3),

$$W_{sat}(E) \sim E^{\beta^*} \quad (L \gg L_x), \tag{3}$$

where $\beta^* = 1.37 \pm 0.10$. Above results predict the first evidence that crystallization process has a new universality class in the nonequilibrium phenomenon such as surface growth. The scaling behavior of the sample of 50 mJ/cm² is similar to that of a-Si. Therefore, we think there are the threshold energy density E_C for crystallization between 50 mJ/cm² and 100 mJ/cm². Furthermore, we must estimate whether the exponent β^* compares to the growth exponent β, i.e., how crystallization by changing the energy density E relates with surface growth as time t [1, 2]. Therefore, a further experiment changing the number of shot at the fixed energy density and analyzing the surface roughness as well should be performed.

We would like to thank Prof. Yamaya for useful discussions. We thank T. Teramoto for experimental support.

References

[1] F. Family and T. Vicsek, eds. *Dynamics of fractal Surfaces*, World Scientific, 1991.

[2] A.-L. Barabasi and H.E. Stanley. *Fractal Concepts in Surface Growth*, Cambridge Universality Press, 1995.

[3] M. Kardar, G. Parisi, and Y.-C. Zang, *Phys. Rev. Lett.*, Vol.56, 1986, pp.889-892.

[4] S.F. Edwards and D.R. Wilkinson, *Pro. R. Soc. London A*, Vol.381, 1982, pp.17-31.

Al'tshuler-Aronov-Spivak Effect and Quantum Chaos in Ballistic Systems

Shiro Kawabata

Department of Applied Physics, Faculty of Engineering, Osaka City University
Sumiyoshi-ku, Osaka 558, JAPAN
E-mail: kawabata@a-phys.eng.osaka-cu.ac.jp

Abstract

The magneto-conductance $G(B)$ in a two-dimensional ballistic system is studied theoretically within the framework of semiclassical scattering theory. The existence of $\Phi_0 / 2$ ($\Phi_0 = hc / e$) oscillation of $G(B)$, analogous to Al'tshuler-Aronov-Spivak effect in disordered metal rings, is theoretically predicted for experimentally-realizable ballistic Aharonov-Bohm billiards. The diagonal-term of the wave-number averaged reflection probability $\delta\mathfrak{R}_D(\Phi)$ is calculated for chaotic and integrable (and mixed) billiards. We find that the difference between chaotic and integrable (and mixed) classical scatterings produces qualitatively different formulas for $\delta\mathfrak{R}_D(\Phi)$ with their behavior determined only by knowledge on the underlying classical dynamics.

Key words: Ballistic transport, weak localization, Al'tshuler-Aronov-Spivak effect, quantum chaos, semiclassical theory

1. Introduction

Advances in nano-fabrication technology have allowed us to study conducting properties of mesoscopic electrical devices with length scale much shorter than both elastic and inelastic mean free paths but larger than Fermi wave length [1]. Recently these ballistic electron devices have attracted much interest as a probe of *"quantum chaos"* [2]. In these systems, various interesting phenomena have been observed in uniform perpendicular magnetic fields, *e.g.*, ballistic weak localization (BWL) effect [3] and the ballistic conductance fluctuation [3]. In the ballistic regime, the semiclassical scattering (SCS) theory is a powerful tool to describe the transport properties. In this paper we shall concern a single ballistic quantum dot in two dimensions (*e.g.*, Sinai billiard) whose structure forms Aharonov-Bohm (AB) geometry and derive analytically the BWL correction term $\delta\mathfrak{R}_D(\Phi)$ to the wave-number averaged reflection probability $\delta\mathfrak{R}_D(\Phi)$ by resorting to the SCS theory, where Φ is the magnetic flux penetrating through the hollow of systems bearing *chaotic* and *integrable* (and *mixed)* phase spaces. We shall show analytically that the average magneto-conductance exhibits $\Phi_0 / 2$ oscillations caused by BWL effect, analogous to Al'tshuler-Aronov-Spivak (AAS) [4] effect in disordered metal rings [5].

2. Semiclassical calculation

In the following, we derive $\delta\mathfrak{R}_D(\Phi)$ separately for chaotic and integrable (and mixed) billiards. First, we consider a two-dimensional ballistic quantum dot which forms an AB geometry (hereafter we call the AB billiard), in uniform normal magnetic field B penetrating only through the hollow. The BWL correction is most easily discussed in terms of the reflection probability R . Therefore, our starting point is the BWL correction term δR to the classical reflection probability R_{cl} $\left(R = \sum_{n,m=1}^{N_M} |r_{n,m}|^2 \sim R_{cl} + \delta R \right)$ with N_M the number of transverse modes,

$$\delta R = \frac{1}{2}\frac{\pi}{kW}\left[\sum_n \sum_{s \neq u} F_{n,n}^{s,u} + \sum_{n \neq m}\sum_{s \neq u} F_{n,m}^{s,u}\right], \tag{1}$$

where

$$F_{n,m}^{s,u}(k) = \sqrt{\bar{A}_s \bar{A}_u} \, \exp\left[ik\,(\tilde{L}_s - \tilde{L}_u) + i\pi v_{s,u}\right]. \tag{2}$$

In eq.(2), s and u label the paths with extremal angles θ and θ', the effective length is $\tilde{L}_s = \tilde{S}_s / k\hbar$, the Maslov indices are included in the phase factor $v_{s,u} = (\bar{\mu}_u - \bar{\mu}_s)/2$, and $\bar{A}_s = (\hbar k/W)\bar{D}_s$. Taking the diagonal approximation, we shall proceed to calculate only the average of $\delta R_D(\Phi)$ over all k which is to be denoted as $\delta\mathfrak{R}_D(\Phi)$:

$$\delta\mathfrak{R}_D(\Phi) = \left\langle \frac{\delta R_D(\Phi)}{kW/\pi} \right\rangle . \tag{3}$$

The only k dependence is involved in the exponential so that the averaging eliminates all paths except for those which satisfy $\tilde{L}_s = \tilde{L}_u$ exactly in eq.(2). A magnetic field through the hollow does not change the classical paths but does change the phase difference between time-reversed paths. Therefore, eq.(3) becomes

$$\delta\mathfrak{R}_D(\Phi) \sim \frac{1}{2}\int_{-1}^{1} d\sin\theta \sum_{s(\theta,\,\pm\theta)} \bar{A}_s \exp\left(4\pi i \frac{\Phi}{\Phi_0} w_s\right), \tag{4}$$

where w_s is the winding number of classical path s. To evaluate the summation over s and integration on θ, we shall reorder the orbits according to the increasing dwelling time T_s . Thus we get

$$\int_{-1}^{1} d\sin\theta \sum_{s(\theta,\,\pm\theta)} \bar{A}_s \sim \int_{T_0}^{\infty} dT\, N(T) \sum_{w=-\infty}^{\infty} P(w) , \tag{5}$$

where $N(T)$ is the classical distribution of dwelling times of a particle in the billiard, $P(w)$ is the winding number distribution of orbits and T_0 is the dwelling time for the shortest winding classical orbits. For chaotic cases (*e.g.*, Sinai billiard), we use $N(T) \sim \exp[-\gamma T]$ for large T [3] . Thus we obtain the result

$$\delta \mathfrak{R}_D(\Phi) \sim \frac{1}{2} \sqrt{\frac{T_0}{2 \, \gamma \alpha}} \left\{ 1 + 2 \sum_{n=1}^{\infty} \exp\left(-n\sqrt{\frac{2T_0\gamma}{\alpha}}\right) \cos\left(4\pi n \frac{\Phi}{\Phi_0}\right) \right\}. \tag{6}$$

In the chaotic case, the main contribution comes from $n=1$ component which oscillates with the period $\Phi_0/2$. On the other hand, for integrable and mixed cases, we use $N(T) \sim T^{-\beta}$ for large T [3] in eq.(5). The same procedure as above yields the qualitatively different result

$$\delta \mathfrak{R}_D(\Phi) \sim \frac{1}{(2\beta-1)\sqrt{2\pi\alpha}T_0^{\beta-1}} \left\{ 1 + 2 \sum_{n=1}^{\infty} F\left(\beta - \frac{1}{2}, \, \beta + \frac{1}{2}; -\frac{n^2}{2\alpha}\right) \cos\left(4\pi n \frac{\Phi}{\Phi_0}\right) \right\}, \tag{7}$$

where F is the hypergeometric function of confluent type. In the case of the integrable and mixed AB billiard the oscillation amplitude decays algebraically, *i.e.*, $F \sim n^{-2\beta-1}$ for large n. Thus, we can see that higher-harmonics components give significant contributions to $\delta \mathfrak{R}_D(\Phi)$ oscillation.

3. Conclusion

We have calculated $\delta \mathfrak{R}_D(\Phi)$ for the ballistic AB billiard. The $\Phi_0/2$ oscillation of $G(B)$, analogous to AAS effect in disordered metal rings, is theoretically predicted for chaotic and integrable (and mixed) billiards. We find that the difference between chaotic and integrable (and mixed) classical scatterings produces qualitatively different formulas for $\delta \mathfrak{R}_D(\Phi)$.

References

[1] C. W. J. Beenakker and H. van Houten, in *Solid State Physics* Vol.44, edited by H. Ehrenreich and D. Turnbull, Academic Press, New York, 1991.

[2] K. Nakamura, *Quantum Chaos: A New Paradigm of Nonlinear Dynamics*, Cambridge University Press, Cambridge, 1993.

[3] H. U. Baranger, R. A. Jalabert and A. D. Stone, *Chaos* Vol.3, 1993, pp.665.

[4] B. L. Al'tshuler, A. G. Aronov and B. Z. Spivak, *Pis'ma Zh. Eksp. Thor. Fiz.* Vol.33, 1981, pp.101 [*JETP Lett.* Vol.33, 1981, pp.94].

[5] S. Kawabata and K. Nakamura, submitted to *J. Phys. Soc. Jpn (Lett.)*.
   S. Kawabata and K. Nakamura, to appear in *Chaos & Quantum Transport*: special issue of *Chaos, Solitons & Fractals*, edited by K. Nakamura, Oxford, Pergamon, 1997.

Bifurcation and dynamical response of current oscillations in semiconductor with NDC

Ken-ichi Oshio

Department of Materials Science, Faculty of Science, Hiroshima University, Higashi-Hiroshima 739, JAPAN

Abstract

We present a simulational study on current oscillations in semiconductors with NDC (Negative Differential Conductivity) which brings nonlinear transport properties such that electric current decreases as increasing bias voltage. An one-dimensional model of gold-doped n-Ge has been proposed, and the computations have been performed under the boundary conditions with d.c. bias. We have found bifurcation diagrams from periodic to chaotic current oscillations, and studied dynamical response of the system to investigate its controllability.

Key words: Spatio-temporal chaos, semiconductor, bifurcation, negative differential conductivity, field-enhanced trapping

1. Introduction

A variety of self-organization and pattern formation in semiconductors, such as high-field domain and high current filament, have been attracted much attention and have been studied by many researchers [1]. Current oscillations in semiconductors are in most cases originated from NDC which occurs due to various physical mechanisms [2]. The recombination instability is one of them, and which is closely related with the field-enhanced trapping effect.
An experimental evidence that the recombination instability gives rise to current oscillations in gold- or cupper-doped n-Ge and high-resistivity GaAs was first reported by Stafeev [3] and Bonch-Bruevich and Kalashnikov [4].
In the 1980s chaotic current oscillations originated from the recombination instability have been reported experimentally and theoretically [5-8]. However, in their theoretical treatments the system has been assumed to be spatially uniform and described by ordinary differential equations. Hence, the computations have been performed only to elucidate the temporal behavior of the system. The current oscillation is closely connected with the successive propagation of high-field domains in the sample. Thus it is very important that partial differential equations are introduced to describe the system in which the spatial degrees of freedom are taken into account, in the aim of this paper.

2. The model for simulation

We have proposed a new model for an one-dimensional system of gold-doped n-Ge in our previous paper, describing time evolution of spatial structure, and showed that three types of modes of operation: the Ohmic (the steady state), the quenched (the high-field domain annihilates before reaching an edge) and the transit-time modes (the high-field domain reaches an edge) have been found. Non-periodic oscillations have been also found in the transition region between the quenched and the transit-time modes [9].

Electric current density $J(x,t)$ is defined as the summation of drift term and diffusion term. Poisson equation is solved simultaneously with the following continuous equations. Time evolution of the occupied trap density $N^-(x,t)$ and the conduction electron density $n(x,t)$ is governed by the continuity equations as follows,

$$\frac{\partial N^-(x,t)}{\partial t} = -\left(\frac{\partial n}{\partial t}\right)_g \quad , \quad \frac{\partial n(x,t)}{\partial t} + \frac{\partial}{\partial x}\left(\frac{J(x,t)}{-e}\right) = \left(\frac{\partial n}{\partial t}\right)_g$$

$$\left(\frac{\partial n}{\partial t}\right)_g = -\alpha_E n(x,t)\{N_t(x) - N^-(x,t)\} + \beta N^-(x,t) \ , \ \alpha_E = \alpha_0 + \alpha_M \exp\left[-\left(\frac{E_{th}}{E(x,t)}\right)^2\right]$$

where $N_t(x)$ is the time-independent total trap density, $E(x,t)$ is the electric field, e is the magnitude of the electronic charge equal to 1.6×10^{-19} [C], α_0 and α_M are the recombination coefficients, and E_{th} is the threshold field for recombination. In the right hand side of the continuity equations, the first term expresses the recombination process in which conduction electrons are trapped by unoccupied traps, and the second term expresses the emission process (the reverse process to the recombination) in which trapped electrons are liberated from the occupied traps due to a thermal and/or an optical excitation. These basic equations are discretized with the aid of the difference procedure which is spatially centered and temporally semi-implicit. The equations of motion are integrated under the boundary conditions with d.c. bias. We have used the values of the system parameters that are considered to be typical of the usual gold-dope n-Ge [9].

The computations were performed to pursue spatio-temporal evolution of the system for several chosen values of the control parameters Φ and β, where Φ is the applied bias voltage and β is the emission coefficient which means the emission of trapped electrons by excitations such as a photoirradiation. The emission is a reverse process against the trapping, so the instability of the system is weaken as increasing β.

3. Results and discussion

We have found a bifurcation route from periodic to chaotic current oscillations at the transition region between the transit-time and the quenched modes, that is the period-doubling route. A global view of dynamics at the transition region is presented through bifurcation diagrams for $\Phi = 0.14$[V] and $7.4 < \beta < 8.4[\times10^7\text{sec}^{-1}]$ (Figs. 1 and 2). Figure 1 and 2 are the cases for increasing and lowering the emission coefficient β, respectively. These diagrams are obtained for each β, by plotting values of minima for the current density time series. Unfamiliar points are observed, e.g., a rectangular-region in Fig. 1. These points originate in spatio-temporal structure during a fundamental period, and those are considered to be a manifestation of precursor phenomenon toward chaotic behaviors. So that may be peculiar bifurcation in a nonlinear system having finite region under time-independent external force. The resemblance between Figs. 1 and 2 means the system has a reversible property for increasing and lowering β. Each simulation has been performed with initial condition which is close to the attractor. Therefore it is expected that the system is not multistable, and each basin for β may be large. Next, to check the controllability of this dynamical system, we have investigated the dynamical response of the system against abrupt change of the parameter β, among period-1(P1), period-2(P2), period-4(P4), and period-8(P8) in the period-doubling cascade (the region where $\beta \geq 8.0[10^7 \text{ sec}^{-1}$ in Fig.1).

Several results are shown in Figs. 3. The control parameter β was changed at the time step $t = 200$. After a control parameter changed, current oscillations at an edge of a sample

settled quickly into the state (the attractor) correspondent to the changed parameter. We have not adjusted the timing for changing β, so initial conditions are selected in a random way. Therefore, this model system has a good controllability for abrupt change of the control parameter β. This controllability can be useful for application, e.g., as an oscillator which generates current oscillations with required period by controlling only one parameter.
The author would like to thank Prof. Shigetoshi Nara for fruitful discussions.

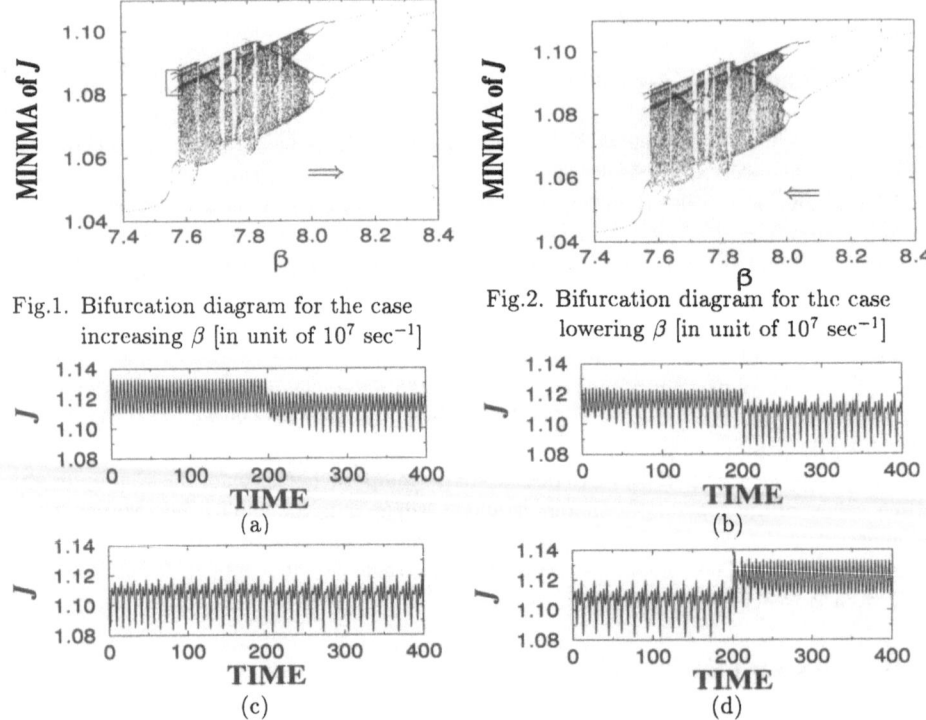

Fig.1. Bifurcation diagram for the case increasing β [in unit of 10^7 sec^{-1}]

Fig.2. Bifurcation diagram for the case lowering β [in unit of 10^7 sec^{-1}]

Fig.3. The current density at an edge of a sample J [in unit of 2.5 A] as a function of time t [in unit of 2.0×10^{-8} sec]. β was changed at $t = 200$. (a) P1→P2 (b) P2→P4 (c) P4→P8 (d) P8→P1.

References

[1] F.-J.Niedernostheide (Ed.): *Nonlinear Dynamics and Pattern Formation in Semiconductors and Devices* (Springer, Berlin, 1995).

[2] E.Schöll: *Nonequilibrium Phase Transitions in Semiconductors* (Springer, Berlin, 1987), and the references cited therein.

[3] V. I. Stafeev: Sov. Phys.-Solid State **5** (1964) 2267.

[4] V. L. Bonch-Bruevich and S. G. Kalashnikov: Sov. Phys.-Solid State **7** (1965) 599.

[5] K. A. Piragas: Sov. Phys.-Semicond. **17** (1983) 652.

[6] S. Bumeliene, J. Požela, and A. Tamaševičius: Phys. Stat. Sol.(b) **134** (1986) K71.

[7] K. Piragas, Yu. Pozhela, A. Tamshyavichyus, and Yu. Ul'bikas: Sov. Phys.-Semicond. **21** (1987) 335.

[8] J. K. Požela, Z. N. Tamaševičiene, A. V. Tamaševičious, J. K. Ulbikas, and G. V. Bandurkina: Phys. Stat. Sol.(a) **110** (1988) 555.

[9] K.Oshio and H.Yahata: J. Phys. Soc. Jpn., **65** (1996) 1490.

Intermittency in the Coupled Chaos Oscillators

Yoshihiro Okagawa, Atsushi Ogawa and Yoshifumi Harada

Deprtment of Applied Physics, Faculty of Engineering, Fukui University, Fukui 910 JAPAN.

Abstract

We performed a simulation of the coupled chaos oscillator which consists of two Chua circuits using a circuit simulator and found an on-off intermittency near the transition to synchronized states. A phase diagram with two parameter space which are a control parameter of the Chua circuit and the strength of the coupling between the circuits was obtained. A complicated route to complete synchronized state was found with increasing the strength of the coupling under the fixed control parameter R around the bifurcation point to chaotic state with a double scroll attractor.

Keywords: on-off intermittency, coupled chaos, Chua circuit, double scroll attractor

1. Introduction

Intermittency has been found in the various nonlinear systems and was classified in the typical types such as typeI, II and III. Rescently a new type of intermittency in coupled chaos systems has attracted great interests and is called on-off[1,2]. The on-off intermittency states consists of burst states (on state) and laminar states (off state). Burst states are localted in narrow time domains and characterized by the self-similar temporal evolution of a state variable. The on-off intermittency was analysed, for example, with a multi plicative noise model[3] and a close relation between this intermittency and the riddled basin was pointed out[4]. Thus on-off intermittency has been recognized universal phenomena in coupled oscillator systems.

Our purpose of this study is to clarify the feature of the temporal evolution in the coupled electric circuit oscillators as a realistic model of the coupled chaos oscillators by using a circuit simulator (PSpice). The circuit is so called Chua circuit[5]. It contains a nonlinear resistor which can be built by an op-amp and linear elements including a resistor. A control parameter is the value of the resistor R (Fig.1).

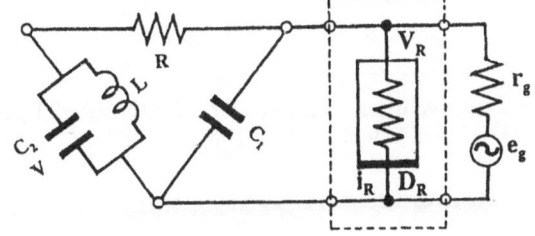

Fig.1 Chua circuit. Inner part of the dashed line is Chua diode.

The Chua circuit provides chaotic oscillation states. The route to chaos is known to be period doubling bifurcation. Characteristic attractors are the periodic attractor (limit cycle), the screw attractor and the double scroll attractor.

For constructing a coupled chaos oscillator system, two Chua circuits are coupled through inductors L_a and L_b with mutual inductance M. The strength of the coupling is defined as $K = M(L_aL_b)^{1/2}$. We investigate various types of coupling by varying the values of R

and K. Here we consider the case two values of R in the oscillators are almost same. In the vicinity of the synchronized state, we found on-off intermittent behavior in the time evolution of the difference of the state variables v_a and v_b in the respective oscillators; $r \equiv v_a - v_b$. Especially we found complicated route to synchronization as increasing the value of K with fixed R around the bifurcation point from the screw attractor state to the double attractor state. Furthermore on-off intermittency state was found not only in the case of chaos - chaos coupling but also in the case of coupling between limit cycle attractor states. It is characteristic feature of our system that the time evolution v_a and v_b are almost periodic in the synchronized states even for the case of chaos - chaos coupling.

2. Simulation

We can consider various types of the way of couplings. Because we treat the case that two values of R in the two oscillators are almost same, the type of coupling can be characterized by their types of attractors. Here we show typical cases such as I):the periodic attractor, II):the screw attractor, III):the double scroll attractor.

Next we briefly show the obtained results. Type I):As increasing the value of K, wave form of the two oscillators become chaotic. Further increasing of K, on-off intermittency is observed for the temporal evolution of r at $K = 0.33$(Fig.2a) and synchroinize with each other at $K = 0.60$ (Fig.2b). Type II:As increasing the strength of the coupling, on-off intermittency is observed at $K = 0.21$ (Fig.3a). The two oscillators synchronize with each other at $K = 0.27$ but on-off intermittency appears again at $K = 0.33$. Complete synchronization occurs at $K = 0.36$ (Fig.3b). Type III:In this state, each system exhibit chaotic oscillation with double scroll attractors for small K. The temporal evolution of r becomes on-off intermittency state at $K = 0.24$ (Fig.4a). Chaotic behavior appears again and transits to complete synchronized state for further increasing of K (Fig.4b). For all cases, when the synchronized state is realized temporal evolution of v_a and v_b in oscillators are almost periodic. Therefore the mechanism of the on-off intermittency is considered to be different from that in ref.[3].

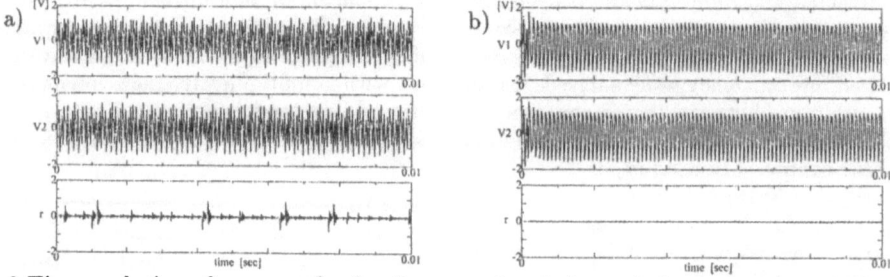

Fig.2 Time evolution of v_a, v_b and r for the case of period - period coupling ($R = 1.70$, $1.69(k\Omega)$). a)$K = 0.33$, b)$K = 0.60$.

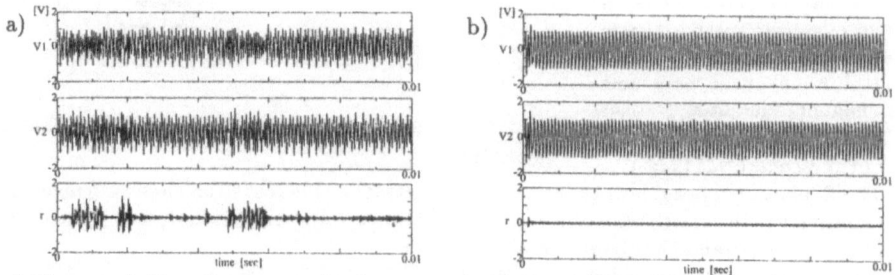

Fig.3 Time evolution of v_a, v_b and r for the case of screw - screw coupling ($R = 1.63$, $1.62(k\Omega)$). a)$K = 0.21$, b)$K = 0.36$.

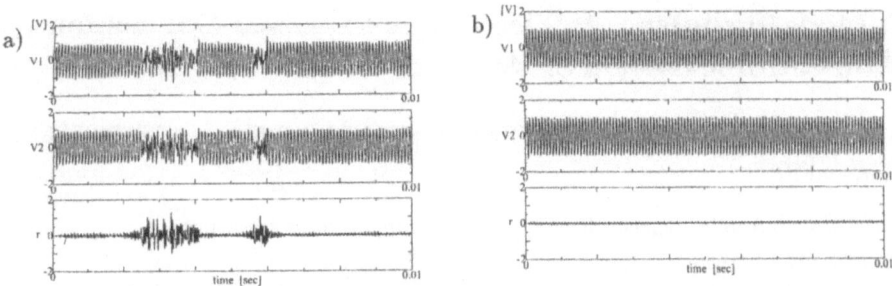

Fig.4 Time evolution of v_a, v_b and r for the case of double scroll - double scroll coupling ($R = 1.58, 1.57(k\Omega)$). a)$K = 0.24$, b)$K = 0.27$.

3. Phase diagram

We construct a phase diagram by changing the values of R and K as shown in Fig.4. We classified the types of temporal variation of r as A:Synchronized state, B:Intermittent state, C:Chaotic state and D:Periodic state. Under the fixed values of R, increasing of K leads to synchronized state. Near the bifurcation point from chaotic state with screw attractors to those with double scroll attractor, two types of synchronization occurs with increasing the value of K. Therefore at the vicinity of these transitions to the synchronized state, on-off intermittency appears.

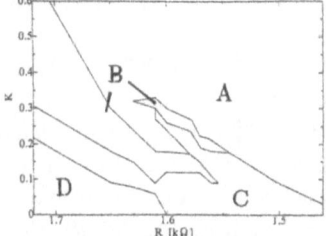

A: Synchronized state
B: Intermittent state
C: Chaotic state
D: Periodic state

Fig.5 The regions for some temporal behavior of r are classified in the phase diagram in (R,K) space as A:Synchronized state, B:Intermittent state, C:Chaotic state and D:Periodic state.

4. Summary

We constructed a coupled chaos oscillator using Chua circuits. We investigated the time evolution of the difference of the state variables in two oscillators and obtained a phase diagram in $R - K$ space. When the synchronization is achieved, the time evolution in oscillators are almost periodic. Around the bifurcation point from the screw attractor state to the double scroll attractor state, the behavior of r is complicated. It needs to analyse the obtained results quantitatively and clarify the mechanism of on-off intermittency. It is easy to extend our coupled chaos oscillator to the case of couplings for large numbers of oscillators and study on this direction is undergoing now.

References

[1] H.Fujisaka, and T.Yamada, Prog. Theor. Phys. **74** (1985) 918; **95** (1986) 1087.
[2] J.F.Heag, N.Platt and S.M.Hammel, Phys. Rev. E49 (1994) 1140.
[3] T.Yamada and H.Fujisaka, Prog. Theor. Phys. **76** (1986) 582.
[4] E.Ott and J.C.Sommerer, Phys. Lett. A188 (1994) 33.
[5] L.O.Chua, M.Komura and T.Matsumoto, IEEE. Circuits Syst. Cas-**33**, (1986) 1072.

Soft-Mode Turbulence in Electrohydrodynamic Convection in Homeotropic System of Nematics

Yoshiki HIDAKA[1], Jong-Hoon HUH[1], Ken-ichi HAYASHI[1], Michael I. TRIBELSKY[2], and Shoichi KAI[1]

1 Department of Applied Science, Faculty of Engineering, Kyushu University,
  Fukuoka 812-81, JAPAN
2 Graduate School of Mathematical Sciences, University of Tokyo,
  3-8-1 Komaba, Meguro-ku, Tokyo 153, JAPAN

Abstract

The experimental investigation of the bifurcation to spatiotemporal chaos was conducted in the electrohydrodynamic instability of homeotropically oriented nematic systems. The electrohydrodynamic instability in the homeotropically oriented nematic system occurred via the Freedericksz transition showing the Goldstone mode due to additional symmetry such as continuous rotational one. It was found that the transition from a spatially uniform rest state to spatiotemporal chaos occurred via the supercritical bifurcation with the softening of the irregular motions toward to the bifurcation point.

Key words: nematics, electrohydrodynamic instability, homeotropic system, Goldstone mode, soft mode

The electrohydrodynamic instability (EHD) of nematic liquid crystals have been investigated intensively as the subject of the studies on dissipative structure and pattern formation in nonequilibrium system. The studies on EHD in the planar oriented nematic liquid crystals with negative dielectric anisotropy have been carried out extensively so far. On the other hand, the EHD in the homeotropic oriented nematic liquid crystals with negative dielectric anisotropy attracts many interests in the pattern formation problem quite recently [1-4]. In the homeotropic system the director **n** is perpendicular to electrodes. Because this alignment state becomes unstable for an electric field parallel to the director **n**, the "buckling" instability that the molecules come to tilt from the initial orientation in arbitrary direction, so-called the Freedericksz transition, occurs for an applied voltage above a threshold V_F [5]. Generally the threshold voltage for convection V_c is higher than for the Freedericksz transition V_F, so the convective instability occurs as the secondary instability.

Figure 1 shows a snapshot of the convective pattern in the homeotropic system. Though this convective pattern is quite disordered, it has a certain characteristic spatial scale due to the convective instability. Furthermore this pattern is not steady but fluctuating with long wavelength and long time scale [4, 6, 7]. In other words, the system indicates the spatiotemporal chaos.

The initial homeotropic alignment state has the continuous symmetry for rotation around an axis perpendicu-

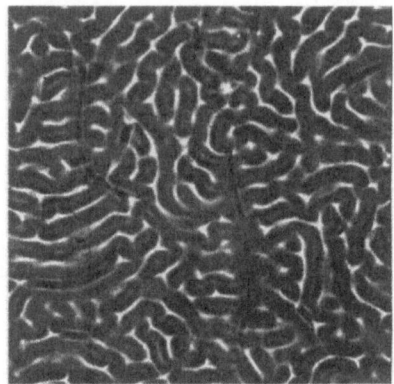

Fig. 1: Convective pattern in homeotropically oriented nematics. Applied voltage is 8.75 V ($\varepsilon = 0.1$) and its frequency is 500 Hz.

lar to the electrode, and spontaneous breaking of the symmetry by the Freedericksz transition that molecules tilt to arbitrary direction brings the Goldstone mode. This mode is fluctuation of tilting molecule in the direction of azimuthal angle around an axis perpendicular to the electrode. The relaxation frequency of the Goldstone mode is equal to zero, that is, this mode is neutrally-stable, because the system after the Freedericksz transition is degenerate with respect to rotation of the director **n** through an arbitrary angle around an axis perpendicular to the electrode.

Recent study by using the one-dimensional modified Swift-Hohenberg equation with an additional continuous group of spatial symmetry obtained the following theoretical results [8, 9]. In the system with an additional continuous symmetry, there exists the Goldstone mode based on the additional continuous symmetry, and all spatially periodic patterns become unstable and spatiotemporal chaos occurs by the nonlinear interaction between the Goldstone mode and conventional "convec-

tion" mode. Furthermore, the bifurcation to spatiotemporal chaos and the transition from spatially uniform rest state to convective one occur simultaneously, that is, a threshold of convective instability corresponds to a bifurcation point from a spatially uniform rest state directly into spatiotemporal chaos. This bifurcation is supercritical one, and the relaxation frequency of the temporal change of the pattern is equal to zero at the bifurcation point, that is, the macroscopic fluctuation by the spatiotemporal chaos is a soft mode. We regard the homeotropic alignment system of nematic liquid crystal as the ideal experimental system with an additional continuous symmetry because the homeotropic system has the additional continuous symmetry of rotation around an axis perpendicular to the electrode, and present the experimental results about the bifurcation to spatiotemporal chaos in the EHD of the homeotropic nematic system in this article.

In order to observe the pattern dynamics for a control parameter ε, a measurement had done by the following procedure. First an applied voltage was raised to 6.00 V between V_F and V_c in order to cause the Freedericksz transition. Subsequently the applied voltage was raised to ε (= $(V^2 - V_c^2) / V_c^2$, $V_c = 8.34$ V) after the enough time (5 min) had passed. Furthermore, after 5 min had passed, the pattern dynamics was taken by the CCD camera and recorded onto a video tape. After the recording of the pattern dynamics the applied voltage was reduced to 0 V in order to erase a hysteresis by the experimental procedure

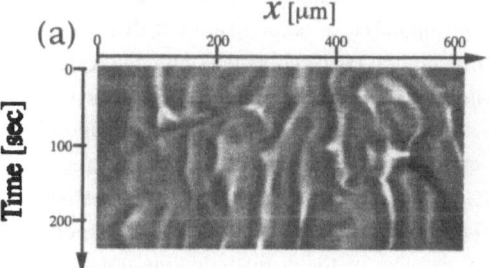

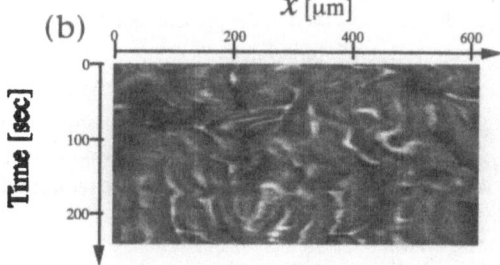

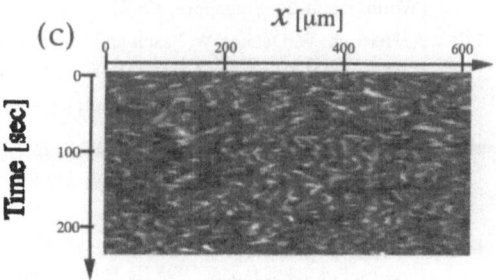

Fig. 2: Spatiotemporal image. The spatial resolution is 1.27 μm / pixel × 512 pixels. The temporal resolution is 1.0 sec / pixel × 256 pixels. (a): ε = 0.10, (b): ε = 0.24, (c): ε = 0.69.

mentioned above. A next experiment was carried out after the system was restored to the original homeotropic alignment. A series of experimental procedure was carried out for some values of ε.

We analysed the experimental data by the following procedure. First the only images on an arbitrary line, defined as x-axis, in the recording image were digitized to one-dimensional profiles of the intensity of 512 pixels with the brightness of 256 gray values and the 256 one-dimensional profiles were sampled at 1.0 sec intervals and placed from the top to the bottom in a chronological order. Figure 2 is a spatiotemporal image made by these procedure. Next auto correlation functions $C(t)$ on time was calculated for all spatial point x from the spatiotemporal image and the average of these functions for x was calculated ($C_{av}(t)$). The correlation time τ was obtained for each ε by fitting the normalized function $C_{av}(t) / C_{av}(0)$ to $\exp(-t / \tau)$ using a least squares program.

Figure 3 shows the dependence of the relaxation frequency τ^{-1} on the control parameter ε. From this result it is found that spatiotemporal chaos occurred at $\varepsilon = 0$, that is, spatiotemporal chaos coincided with convective instability. Moreover, this result indicate that supercritical bifurcation into spatiotemporal chaos occurred and macroscopic fluctuations softened as the control parameter approached into the bifurcation point of the spatiotemporal chaos, that is, the critical-slowing down occurred. Therefore we named the spatiotemporal disorder observed in the present experiment as the soft mode turbulence.

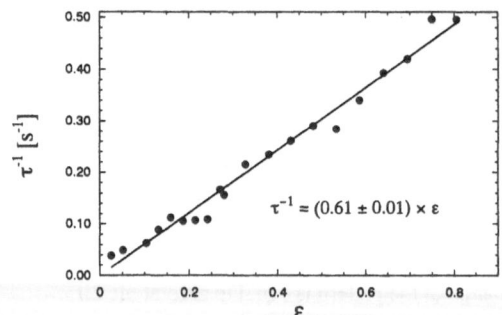

Fig. 3: Dependence of relaxation frequency τ^{-1} on the control parameter ε.

We conducted the experimental investigation of the bifurcation to spatiotemporal chaos in the electrohydrodynamic instability of homeotropically oriented nematic systems. It was found that the transition from a spatially uniform rest state to spatiotemporal chaos occurred via the supercritical bifurcation with the softening of the irregular motions toward to the bifurcation point.

References

[1]   L. Kramer, A. Hertrich and A. Pesch, *Pattern Formation in Complex Dissipative Systems*, ed. S. Kai (World Scientific, Singapore, 1992) pp.238.

[2]   A. Hertrich, W. Decker, W. Pesch and L. Kramer, J. Phys. (Paris) II **2**, 1915 (1992).

[3]   K. Hayashi, Y. Hidaka and S. Kai, *Proc. of 1st Tohwa Univ. Int. Meeting on Statistical Physics*, Fukuoka, Japan, 1995 [in Bussei Kenkyu (Kyoto) **66**, 504 (1996)].

[4]   R. Richter, N. Klopper, A. Hertrich and A. Buka, Europhys. Lett. **30**, 37 (1995).

[5]   P. G. de Gennes and J. Prost, *The Physics of Liquid Crystals* (Oxford University Press, New York, 1993) 2nd ed.

[6]   M. I. Tribelsky, K. Hayashi, Y. Hidaka and S. Kai, *Proc. of 1st Tohwa Univ. Int. Meeting on Statistical Physics*, Fukuoka, Japan, 1995 [in Bussei Kenkyu (Kyoto) **66**, 592 (1996)].

[7]   Y. Hidaka, K. Hayashi and S. Kai, J. Jpn. Soc. Fluid Mech. **15**, 163 (1996) (in Japanese).

[8]   M. I. Tribelsky and K. Tsuboi, Phys. Rev. Lett. Phys. **76**, 1631 (1996).

[9]   M. I. Tribelsky, to be published in Int. J. Bifurcation and Chaos.

Temporal Order and Dynamical critical phenomena in Sliding Charge-Density Wave Systems

Ken-ichi Matsuda, Satoshi Tanda, Yoshitoshi Okajima, and Kazuhiko Yamaya

Dept. of Applied Phys., Hokkaido Univ., Sapporo 060, Japan

Abstract

We investigated the transport properties of sliding charge-density wave (CDW) systems in the presence of impurities at 77.3K. Near the threshold bias voltage V_T. current carried by CDW (I_{CDW}) was scaled by reduced bias voltage $(V - V_T)/V_T$ with exponent $\zeta = 1.1 \pm 0.1$. This scaling behavior gives good agreement with a theoretical picture proposed by Fisher. Moreover, we found that the amplitude of narrow-band noise (NBN) became minimal at a critical voltage V_C. This implies the non-equilibrium phase transition between a plastic flow phase and a temporally periodic moving solid phase. We confirmed the existence of three dynamical phases in non-equilibrium steady state.

Key words: Charge-density wave, Dynamical critical phenomena, Noise spectrum, Temporal order

1 Introduction

The influence of randomly quenched impurities on the transport properties in charge-density wave (CDW) systems has been a subject of intense study[1]. One reason for interest in sliding CDW's has been a major focus on the dynamical critical phenomena of many-degree-of-freedom systems, such as vortices in the mixed state of type II superconductors and the sandpile-avalanches. The impurities destroy the spatial long range order and pin the CDW's. However, the systems depin and become mobile in the presence of driving force provided by an electric field. One important consequence of this competition between the random pinning and the elasticity of the CDW itself is the existence of a depinning transition. Since there is an analogy between the sandpile-avalanches and the collective dynamics of the CDW's, the process of depinning can be interpreted as a dynamical phase transition between the pinned(creep phase) and sliding states (plastic flow phase) of the CDW[2]. The order parameter for the depinning transition is the average CDW velocity v_{CDW} and it is observed experimentally as a CDW current I_{CDW}. According to the theoretical studies, the evidence for the existence of dynamical phase transition is a scaling behavior of I_{CDW} near the threshold voltage V_T. Another consequence is the existence of a non-equilibrium phase transition between plastic flow phase and moving solid phase in the sliding states of the CDW's. Just above the threshold voltage, the CDW's are in the plastic flow phase. With increasing bias voltage. the impurities are less effective and the CDW's become into the moving solid phase. In the moving solid phase, despite the absence of spatial periodicity, the CDW's have long range temporal correlation. The experimental signature of such a temporally ordered state correspond to the observation of a narrow band noise (NBN).

In this paper, we report the experimental evidence for the dynamical critical phenomena of

CDW's in non-equilibrium steady state at 77.3K. At the depinning transition, we observed a scaling behavior of I_{CDW}. In the sliding states, the fundamental NBN peak moved to higher frequency with increasing bias voltage. The power of NBN peak strongly depends on bias voltage and became minimal at $V_c = 64.5$ mV. We confirmed the existence of three dynamical phases of CDW's in the presence of bias voltage.

2   Experimental

We used o-TaS$_3$ crystals in this study. Typical crystal dimensions were 5mm $\times$ 10μm $\times$ 1μm, where the long dimension corresponds to the chain axis. The sample was placed on a glass substrate and electrical contacts were made by evaporating gold. The length for each of the intercontact arms was $\sim$0.5 mm. We measured $I - V$ characteristics and voltage fluctuations with 4 probe configuration at 77.3K. In the voltage fluctuation measurement, the signal was amplified with an ultra low noise amplifier and fed to a digital spectrum analyzer. Each of spectrum data was obtained by averaging 8196 data.

3   Results and discussions

3.1   Nonlinear conductivity

Figure 1 shows the scaling behavior of the CDW current I_{CDW} near the threshold voltage V_T. The horizontal axis represents reduced bias voltage $(V - V_T)/V_T$, where V_T is 34.5 mV for this sample. The solid line shown in figure indicates a power law form

$$I_{CDW} \sim \left(\frac{V - V_T}{V_T}\right)^{\zeta} \tag{1}$$

where ζ is a dynamical critical exponent. One can find that the data are well fitted in the range of $(V - V_T)/V_T < 0.9$ with $\zeta = 1.1 \pm 0.1$. This result gives good agreement with theoretical prediction. Thus the depinning transition can be interpreted as a dymanical phase transition. Moreover, the value of exponent which obtained from our experiment is close to that of other CDW's [3][4]. This implies that quasi-one dimensional CDW's belong to the same universality class. In the range of $(V - V_T)/V_T > 0.9$, however, the scaling behavior vanishes and I_{CDW} increases rapidly. It indicates that the transport properties in CDW's change and become a temporally periodic moving solid phase.

3.2   Noise spectra

Voltage fluctuation measurement was performed in o-TaS$_3$ at 77.3K. Below V_T, there is no evidence for either Broad-band noise (BBN) or Narrow-band noise (NBN). For $V > V_T$, however, both NBN and BBN were observed. The experiment was actually performed with nearly fixed current and thus V represents the time-averaged dc value. With increasing the bias voltage, fundamental spectrum of NBN gradually appeared and its frequency f_0 shifted to higher frequency. Figure 2 shows that the bias dependence of the NBN spectrum. Each of profiles is normalized by half-power bandwidth Δf and thus the horizontal axis represents a normalized frequency f_{norm} defined as $f_{norm} = (f - f_0)/\Delta f$. With increasing the bias voltage, the power of NBN $S(f_0)$ suddenly decreases and takes a minimum value at $V_c = 64.5$ mV. For $V > V_c$, the sharp profile of NBN was observed and it indicates that the systems become into temporally ordered state. In the recent theoretical study [1], The existence of non-equilibrium phase transition in sliding CDW's was predicted. The experimental signature of such phase transition is scaling behavior of $S(f_0)$ near the critical bias voltage V_c.

The relationship between $S(f_0)$ and bias voltage are shown in figure 3. At $V_c = 64.5$ mV, $S(f_0)$

takes minimum value. Although scaling behavior of $S(f_0)$ is not observed, this result implies the existence of non-equilibrium phase transition between plastic flow and Moving solid. This is the first experimental observation.

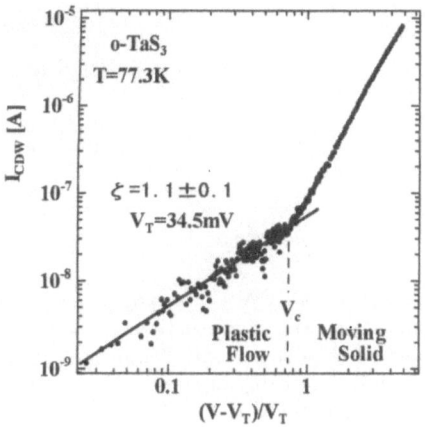

Fig.1 : Scaling plot of I_{CDW} obtained from the dc $I - V$ curves in o-TaS$_3$. The experiment was performed at 77.3K. The critical exponent obtained from the solid-line shown in figure is $\zeta = 1.1 \pm 0.1$, $V_T = 34.5$ mV for this sample.

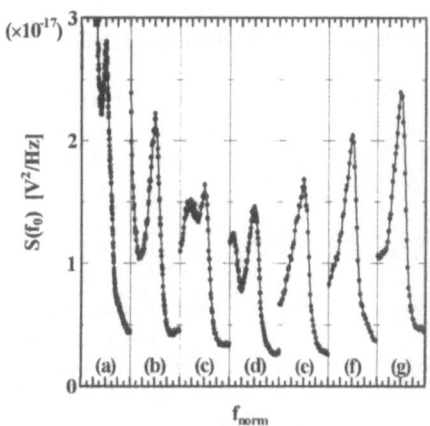

Fig.2 : The bias dependence of the spectrum of NBN in o-TaS$_3$ at T=77.3K. the parameters of each data are (a)$V = 52.52$ mV, $f_0 = 405$ Hz (b) 55.93 mV. 875 Hz (c) 60.45mV, 2150 Hz (d) 65.98mV. 4700 Hz (e) 73.26mV, 10625 Hz (f) 83.97mV. 24125 Hz (g) 94.07mV, 40375 Hz.

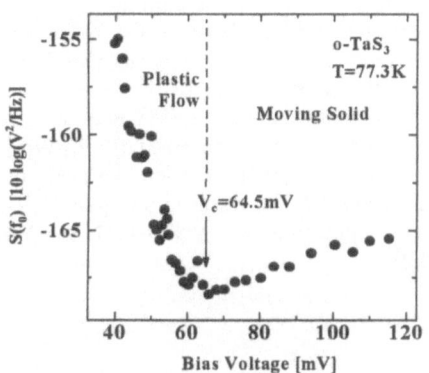

Fig.3 : The relationship between $S(f_0)$ and bias voltage at T = 77.3 K. $S(f_0)$ takes minimum value at $V_c = 64.5$mV.

We would like to thank Prof. T. Sambongi for providing samples and useful discussions.

References

[1] L. Balents, and M.P.A. Fisher, *Physical Review Letters*, Vol.75, 1995, pp. 4270–4273
[2] D.S. Fisher, *Physical Review B*, Vol.31, 1985, pp. 1396–1427
[3] S. Bhattacharya, M.J. Higgins, and J.P. Stokes, *Physical Review Letters*, Vol.63, 1989, pp. 1503–1506
[4] J. Peinke, A. Kittel, and J. Dumas, *Europhysics Letters*, Vol.18, 1992, pp. 125–131

Competing Instabilities and Localized Traveling Wave in Binary Fluid Mixture

Lizhong Ning[1], Yoshifumi Harada[1], and Hideo Yahata[2]

1. Dept. of Applied Physics, Faculty of Engineering, Fukui Univ., Bunkyo 3-9-11, Fukui 910, JAPAN
2. Dept. of Materials Science, Faculty of Science, Hiroshima Univ., Higashi-Hiroshima 739, JAPAN

Abstract

Dynamics of defects and double localized traveling waves (DLTW) in a large-aspect-ratio L =46 and in a strongly nonlinear regime with a separation ratio $\psi = -0.47$ have been studied by the simulation of full hydrodynamic equations and by the experiment. We find that a defect structure can be formed by two different processes, and remarkably influences the structure of the DLTW. The length of the DLTW is a function of the reduced Rayleigh number r.

Key words: Defect, double localized traveling wave (DLTW), numerical simulation

1. Introduction

Experiments on traveling wave (TW) convection in a binary fluid layer heated from below have revealed a wealth of TW convection patterns. Harada et al. at first observed the DLTW state slightly above the saddle-node point on the upper branch of a bifurcation curve in a long slot channel, and found that some intrinsic connection exists between the defect and the DLTW[1, 2]. Amplitude equations have been used to understand qualitatively the DLTW[3]. On the numerical simulation side of full hydrodynamic equations, Barten et al. have quantitively simulated the LTW[4], but have not found the spatiotemporal defect and the DLTW. Using the system with rigid lateral boundary conditions, we have obtained the LTW[5], the DLTW[6] and the spatiotemporal defect[7]. In this paper, we mainly present the experimental and simulational results on the spatiotemporal defect and DLTW states in a large-aspect-ratio $L = 46$ and in a strongly nonlinear regime with the separation ratio $\psi = -0.47$, the Prandtl number $Pr = 13.8$, and the Lewis number $Le = 0.01$.

2. Governing equations of system

Using the Oberbeck-Boussinesq approximations and neglecting the Dufour effect, the two-dimensional governing equations of the system can be written as

$$\partial_t \mathbf{u} + (\mathbf{u} \cdot \nabla)\mathbf{u} = \nabla^2 \mathbf{u} - \nabla(p/\rho_0) + (1/Pr)[(1 + \psi)\theta - \psi\eta]\mathbf{e_z} \tag{1}$$

$$\nabla \cdot \mathbf{u} = 0 \tag{2}$$

$$\partial_t \theta + (\mathbf{u} \cdot \nabla)\theta = (1/Pr)\nabla^2\theta + Ru_z \tag{3}$$

$$\partial_t \eta + (\mathbf{u} \cdot \nabla)\eta = (1/Pr)(\nabla^2\theta + Le\nabla^2\eta) \tag{4}$$

where $\mathbf{u}(u_x, u_z)$, θ, η, p and t denote the dimensionless velocity, temperature, concentration current, pressure and time respectively, R is the Rayleigh number. The method for solving the governing equations and the initial conditions are described in Refs. 6 and 7.

3. TW state containing defects and its structure

Figure 1 shows the transient defect structure above the onset of convection obtained by the numerical simulation. The system evolves to the defect state in a nonlinear stage after a linear stage[7]. In this stage, defects appear continuously as shown in two-dimensional space-time diagram Fig.1 in which a wave number adjusts itself with time, and the TW containing the defects dominates the confined region. In the experiment, the system first passes through a similar linear stage, then evolves to the defect state, in which the effect of nonlinearity leads to decreasing in the speed of the TW, as shown in Fig.2. However, the space-time structure in Fig.1 is more complicated than that in the present experiment, since the TW not only contains the source defect but also possesses the sink defect as observed in some other experiments.

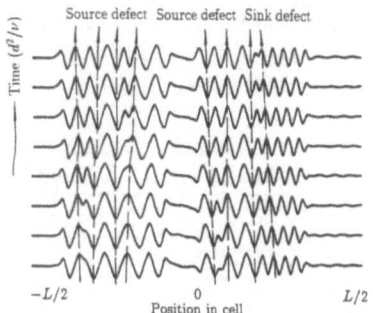

Fig.1 Transient defect structure at $r = 2.1$, time difference Δt between two lines is $1.25 \cdot d^2 / \nu$.

Fig.2 Transient space-time structure at $r = 6.53$, each line shows a position of a downflow.

In the experiment, a stable defect state shown in Fig.1 in Ref.1 appears, and the number of the defects changes as a function of r[1] over a range of r. Figure 3 shows a stable sink defect obtained in the numerical simulation. Comparing the defect structure between Figs.1 and 3, we find that the formation of the defect can be categorized into the two processes: (1) A pair of rolls is generated or annihilated between two rolls, say, between two broken lines shown in Fig.1. This mechanism of formation has also been observed in the experiment of Kolodner[8]. (2) One roll abruptly generates the phase slip with half a period, at the same time, one roll is annihilated or generated on each of its side, as shown in Fig.3. This agrees with the simulational result for a source defect in a weakly nonlinear regime with $L = 12$. It should be noted that all the complicated defect structure such as dislocations and grain boundaries[7,8,9] can be decomposed into two types of basic structures: a source defect and a sink defect.

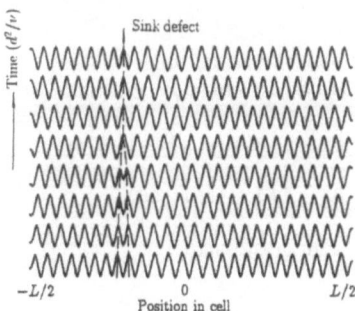

Fig.3 Sink defect at $r = 2.1$, time difference Δt between two lines is $1.25 \cdot d^2 / \nu$.

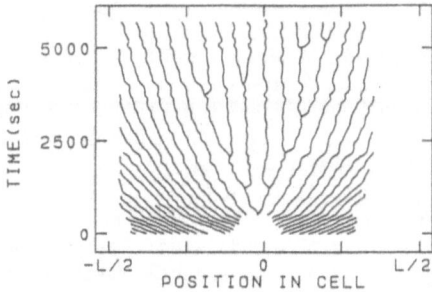

Fig.4 Experimental DLTW at $r = 5.35$, each line shows a position of a downflow.

4. Double localized traveling wave state and its behavior

With further decreasing r, the system evolves to the DLTW state shown in Fig.4 from the TW
containing the defect. Experiment shows that the stable region of the DLTW depends on how
it is prepared experimentally, and the length of the DLTW is a function of r over the region in
which the DLTW exists. The length L_1 of the TW near the left end wall shown in Fig.5 ranges
from $0.08L$ to $0.45L$, and the one near the right end wall also shows a similar dependence.
Figure 6 shows time evolution of the temperature field in the DLTW reached after the passage
of a long time from the final state in Fig.3 by changing r to 1.35 in the numerical simulation.
Numerical simulation has thus shown the existence of the DLTW, and L_1 is nearly $0.11L$ and
the DLTW exists just above the saddle-node point similar to experimental observation. The
difference of the moving direction of the TW near the right end wall between Figs.4 and 6
originates from the defect structures shown in Fig.1 in Ref.1 and Fig.3 of this paper, and shows
that the defect structure remarkably influences the structure of the DLTW. The difference
between the values of r over which the DLTW exists has been discussed in Ref.7.

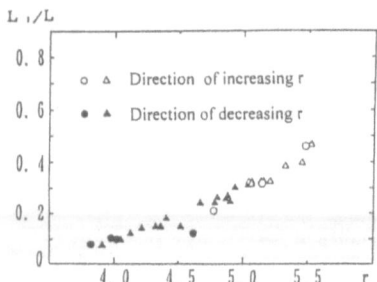
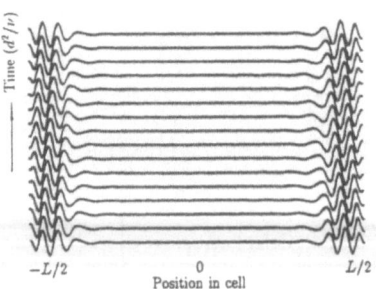

Fig.5 r dependence of DLTW length

Fig.6 Simulated DLTW at $r = 1.35$, time difference
Δt between two lines is $1.25 \cdot d^2/\nu$.

5. Conclusion

Using the simulation of full hydrodynamic equations, we obtained the solution of the DLTW observed in the experiment and found that the defect can be formed by the two different processes.
A comparison between the simulation and the experiment shows that the defect remarkably
influences the structure of the DLTW. Therefore, in order to understand the formation of the
DLTW, it be useful to further study in detail the dynamics of the defect.

References

[1] Y. Harada, Y. Masuno, and K. Sugihara, Vistas in Astronomy **37** (1993), 107.
[2] Y. Harada, Y. Masuno, K. Sugihara, K. Nomura, and H. Yahata, in *Pattern Formation in Complex
Dissipative Systems*, ed. S. Kai (World Scientific, 1992), p314.
[3]A. Otake, A. Ogawa and Y. Harada, in *Dynamical Systems and Chaos*, ed. Y. Aizawa (World
Scientific, 1994), p.234.
[4] W. Barten, M. Lucke, M. Kamps, and R. Schmitz, Phys. Rev. **E51** (1995), 5662.
[5] H. Yahata, Prog. Theor. Phys. **85** (1991), 933.
[6] L. Z. Ning, Y. Harada, and H. Yahata, Prog. Theor. Phys. **96** (1996), 668.
[7] L. Z. Ning, Y. Harada, and H. Yahata, (1996), preprint.
[8] P. Kolodner, Phys. Rev. **A46** (1992), 6452.
[9] D. Bensimon, P. Kolodner, C. M. Surko, H. Willisams and V. Croquette, J. Fluid Mech. **217**
(1990), 441.

Experimental Evidence of Marangoni Convection Rolls in Belousov-Zhabotinsky Reaction

Osamu Inomoto [1,2], Akemi Sakaguchi [1], and Shoichi Kai [1]

[1] Department of Applied Physics, Kyushu University 36, Fukuoka 812-81, Japan
[2] Research Fellow of the Japan Society for the Promotion of Science

Abstract

We report experimental evidence of convection rolls induced by the Marangoni effects in the bigwave (hydrochemical soliton) in the Belousov-Zhabotinsky (BZ) reaction. It is caused by spatial inhomogeneity of surface tension between the front and the back of the chemical waves on the solution layer. We show concentration distributions of ferriin obtained from the two different waves such as the bigwave and the normal trigger wave. Surface tension distributions chemically originated are examined to explain the mechanism of flows. It is clear that the temperature difference on the surface due to the reaction must be taken into account.

Key words: Belousov-Zhabotinsky reaction, Marangoni convection, bigwave, hydrochemical soliton

1. Introduction

BZ reaction is a well known example of nonequilibrium dissipative systems [1]. In this reaction system, self-organized patterns (the so called *chemical waves*) are observed in shallow layers. These patterns are emerged by coupling of nonlinear reaction and diffusion mechanism under some externally suitable conditions.

We have reported the curious chemical wave, called *the bigwave* (*hydrochemical soliton*), accompanied by large hydrodynamic flow in the solution layer [2-5]. It has been clarified that the flow is caused by the surface tension instability (the Marangoni instability) with nonlinear coupling between flows and chemical reactions.

The origin of the surface tension changes however has not been well understood yet, while we propose two possibilities such as the change of ferriin to ferroin concentration ratio and the temperature elevation by the exothermic reaction [3].

Recently, M.Boeckmann *et al.* denied the possibility of exothermic origin for the induced flow from their experimental evidence in spiral waves [6]. However it is not the case in the bigwave because the bigwave appears only under strongly excited states. Therefore still two possibilities are alive.

In this work, in order to understand the real origin, at first we only focus on the concentration changes of ferroin which is a kind of surfactant. Measurements of surface temperature for the bigwaves are now in preparation and not available yet. We compare concentration distribution of ferriin for the bigwave to that for a normal trigger wave which is not accompanied by hydrodynamic flows.

2. Experimental

We prepared batch layers of BZ reaction solutions which had long induction periods (∼10min). The compositions of the solutions were malonic acid, sodium bromate, sodium bromide, ferroin, and sulfuric acid, of which concentrations were 95mM, 340mM, 48mM, 3.5mM, and 378mM, respectively.

The concentration distributions of ferriin were measured using 488nm of Argon laser (NEC GLS-3200) and a charge-coupled device camera (Panasonic NV-X100). According to the Lambert-Beer's law, light transmittance can be converted into the ferriin concentration at this wavelength. The extinction coefficient of the ferroin at 488nm is 10090 ± 285 [$M^{-1}cm^{-1}$], and the absorption by the ferriin is neglected [7]. As the quantitative analysis of the dependence of surface tension on the ferroin (ferriin) concentration is already reported [8], the relation between the magnitude of surface tension and the absorption can be calculated. Thus we obtain the value of surface tension from light transmittance.

3. Results and discussions
3.1 Ferroin concentration distributions

Fig.1 shows ferriin concentration profiles along x, the axis of the propagation of the chemical waves. BW and NW in the figure mean the bigwave and the normal trigger wave respectively. Both waves propagate from left to right, and $x = 0$ indicates the position of the fronts.

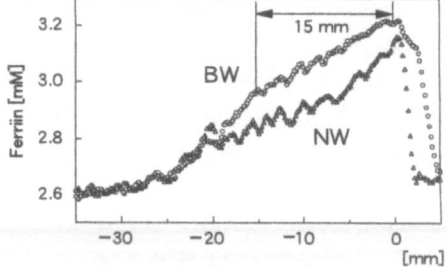

Fig.1 Ferriin concentration distributions along to the direction of wave propagation.
BW: bigwave, NW: normal trigger wave
Each waves propagate from left to right and x = 0 is the position of both chemical wave front. Here the following relation is established: [ferriin] = 3.5mM - [ferroin].

The characteristic difference of the ferriin concentration distributions between the two waves is clear. The profile of BW has a specific shape where the linear decrease of the concentration is observed from just behind the chemical wave front to 15mm. This area must be correspond to the region where a convection roll occurs due to the Marangoni effect. On the other hand in NW no such a profile is observed and instead simple decay from the front is observed of which tendency has been already reported [7].

According to the result, no remarkable difference of the concentration gradients between both waves is observed. This means that the chemical origin of BW is not only by the surface tension gradient due to the ferriin concentration. The surface temperature by the exothermic reaction therefore must be also taken into account in the surface tension change.

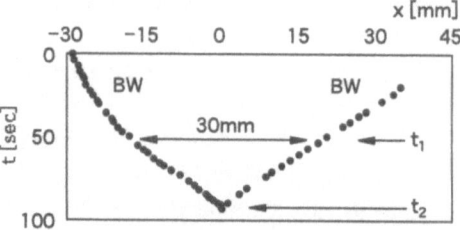

Fig.2 Temporal evolution of two interacting bigwaves. Traces of the wave fronts are plotted in a spatiotemporal map. They decrease their propagating speed at $t = t_1$ (at this moment the distance of the fronts is 30mm) and vanish at $t = t_2$ by collision.

3.2 Collision of two bigwaves and a front convection roll

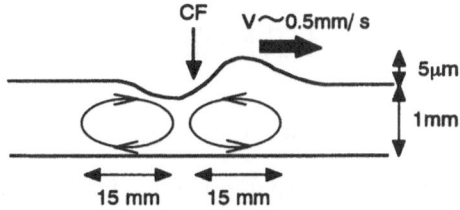

Fig.2 shows an interaction of two bigwaves in spatiotemporal map. The each ones "recognized" the other at $t = t_1$ before the "chemical collision" occurs at $t = t_2$ and their propagating velocities start to decrease. Therefore waves are not totally independent each other. The distance between two wave-fronts at $t = t_1$ is about 30mm of which value corresponds to the twice of the size of the convective roll described in 3.1.

Fig.3 Schematic drawing of convection flow in a bigwave. CF and v denote the position of the chemical wave front and the vector of its propagation, respectively.

Based on the results here, the hydrodynamic structure as described in Fig.3 can be expected. The new outlook is reasonable to the previous knowledge [2-5]. Precise and direct observations of the hydrodynamic flow by using laser Doppler velocimeter are now in progress.

4. Conclusion

We described in this short article the evidence of a pair of convection rolls in the chemical wave front of the bigwave.

As Boeckmann et al. pointed out, the effect of the concentration change of the chemicals may be much larger than that of the temperature change in order to lead the flow instability in BZ reactions [6]. However, the result of Fig.1 asserts that peculiarities of the bigwave cannot be explained only by the ferriin (ferroin) concentration distribution, since it does not differ enough from that of normal trigger wave. Obviously both the gradients affect to the instability and the bigwave may involve huge heat of reaction different from normal trigger waves. To make our argument clear, measurement of spatial temperature distribution on the surface is necessary. We will report more careful and systematic results in near future.

5. Acknowledgements

We would like to acknowledge valuable discussions with Dr.T.Yamaguchi (NIMC). This work was partially supported by the Grants-in-Aid for scientific research from the Ministry of Education, Science and Culture, Japan.

6. References

[1] R.J.Field and M.Burger (Ed.), *Oscillations and Traveling Waves in Chemical Systems*, John Wiley & Sons, 1985.

[2] H.Miike et al., Phys.Rev.E48 (1993) 1627.

[3] S.Kai and H.Miike, Physica A204 (1994) 346.

[4] S.Kai et al., Physica D84 (1995) 269.

[5] O.Inomoto et al., J.Phys.Soc.Jpn.64 (1995) 3602.; Int'l. J.Bifur.Chaos (1997) in press.

[6] M.Boeckmann et al., Phys.Rev.E 53 (1996) 5498.

[7] P.M.Wood and J.Ross, J.Chem.Phys.82 (1985) 1924.

[8] K.Yoshikawa et al., Chem.Phys.Lett.211 (1993) 211.

Localized Traveling Wave Front in Binary Fluid Convection

Masashi Nomura, Atsushi Ogawa and Yoshifumi Harada

Department of Applied Physics, Faculty of Engineering, Fukui University, Fukui 910 JAPAN.

Abstract

We performed an experiment on Rayleigh Bénard convection in a binary fluid in the strong non-linear regime in a slot channel. Double localized traveling wave (DLTW) state was realized near the saddle node point. We have measured the local dynamics at the fronts which separate a convection region and a conduction region in the DLTW and found the chaotic dynamics at the fronts. We suggest that the fronts in DLTW as a cohernt structure are stabilized by the chaotic dynamics.

Keywords: Rayleigh Bénard convection, binary fluid, slot channel, double localized traveling wave, chaos

1. Introduction

Rayleigh Bénard convection in binary fluid have been one of the fascinating phenomena in non-equilibrium systems because their rich variety of spatio-temporal behavior[1]. The most striking phenomena in binary fluid convection is localized structures in the form of confined traveling waves *i.e. localized traveling wave* (LTW). Because these localized states are characteristic structure in non-equilibrium systems, it has been recognized that the importance of studying the structure and the formation mechanism of them.

A great deal of efforts have been devoted to the study on LTW in the weakly non-linear regime, because this case is theoretically tractable. From the experiments in annular cells, LTW has been recognized an intrinsic feature of the unbounded system and not caused by the interaction of the TW's with the end walls of the cell[2]. Kolodner's experiments showed the LTW drift slowly as pulses in the same direction as the phase velocity of the underlying TW[3]. The feature of LTW is reproduced by numerical integration of the full Navier Stokes equations in two dimensions[4] and by complex Ginzburg Landau equations coupled to slow mean concentration field[5]. LTW in the strongly non-linear region are clearly different from that in the weakly non-linear regime and is not understood sufficiently[6].

On the other hand, the experiment of the Rayleigh Bénard convection in pure fluid using slot channel cell one of the excellent model for one-dimensional spatio-temporal pattern formation[7]. When the width of the cell is very narrow, additional strong non-linear effects arise in convective behavior. We have studied the strong non-linear Rayleigh Bénard convection in binary fluids in the slot channel and found new types of LTW such as double localized traveling wave (DLTW)[8]. The formation process and the stability of DLTW are quite different from the conventional types of LTW. We have predicted the existence of the chaotic dynamics at the fronts in DLTW which separate a convection region and a conduction region. Our purpose in this paper is to report the experimental results on the local dynamics of the fronts in DLTW. By the measurement of the intensity of the shadow graph signal as a function of time, we have found that the chaotic motion exist. We believe that this chaotic behavior is quite important for the mechanism of stabilization of the fronts in DLTW.

2. Experiment

2.1 Experimental apparatus and procedure

The fluid used in oue experiment is 8wt% solution of ethyl alcohol in water. The side and end walls of the cell are defined by plexi-glass and optical glass respectively. The inner width between side walls is $W = 0.30$ cm and the inner distance between the end walls is $L = 24$ cm. The height of the cell is $d = 0.52$ cm. Therefore the aspect ratio of the cell is $1 : \frac{W}{d} : \frac{L}{d} = 1 : 0.6 : 46.2$ and is classified as a slot channel cell. The bottom plate was heated from below and the top plate was cooled from above by circulating water which temperature was fixed at $10.18°$C. The temperature difference ΔT applied across the cell was set up by the control of the vertical heat flux. Shadow graph technique was used for flow visualization.

The separation ratio is $\psi = -0.46$ which corresponds to the strong non-linear region and it leads to the strong sub-critical bifurcation. The Prandtl number and the Lewis number are $Pr = 13.6$ and $Le = 0.01$ respectively.

2.2 Experimental results

In this paper we use the reduced Rayleigh number defined as $r = \frac{R}{R_{co}}$ in which R is usual Rayleigh number and $R_{co}(= 1707.8)$ is the threshold value for the onset of convection in pure fluid. As increasing the temperature difference, the instability occurred at $\Delta T = 9.96°$C. It corresponds to $r = 7.32$ which is denoted as r_{co}. This value is quite large compared to the predicted one from the linear stability analysis due to the very narrow width cell. After the transient process with sub-critical bifurcation, a uniform TW state was realized.

As increasing r, the phase velocity of the TW gradually decreased and transited to the stationary overturning convection (SOC) state. As decreasing the value of r from SOC state, TW states are realized again. Further decreasing of r, defects were induced in TW state and DLTW state was formed as shown in Fig.1 near the saddle node point. It is characteristic feature of DLTW that two TW's have phase velocities and wave lengths different from each other[8]. The boundaries between fast localized traveling wave and conduction state is a trailing edge front and that between slow localized traveling wave and conduction state is a leading edge front.

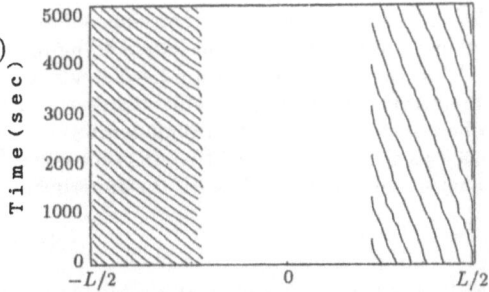

Fig.1 Spatio-temporal plot of the DLTW state at $r = 4.85$. The solid lines correspond to the posions of the down flow.

This DLTW is triggered by the appearance of defects in the TW state. Defects are created due to the spatial inhomogeneity of the TW state. Such inhomogeneity seems to be caused by the feature of the long and narrow cell. The detailed successive transient process of convection patterns in the slot channel cell will be reported elsewhere.

To investigate the dynamics at the fronts, we observed the intensity of the shadow graph signal as a function of time. The wave form reflects the spatial structure of the density field of the TW and is obviously fluctuating as shown in Fig.2 a). From the time series, we constructed an attractor in the three dimensional space $x(t)$, $x(t+\tau)$ and $x(t+2\tau)$ where τ is a delayed time. Fig.2 b) is the attractor constructed from the time series at the leading edge front. This attractor reflects the aspect of an edge chaos. To confirm such a weak chaotic behavior quantitatively, we calculate the value of Lyapunov exponent λ and obtained the positive value $\lambda = 8.60 \times 10^{-4}$. This chaotic behavior was observed also at the trailing edge front with smaller value of λ. On the other hand, the values of λ in the region of TW are negative.

From these observations, we conclude that the chaotic dynamics exist at the fronts of DLTW. We suggest that the stability of the fronts in DLTW in slot channel cells is closely related to these chaotic dynamics.

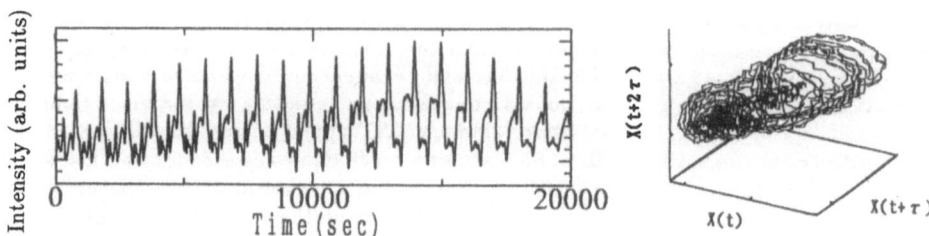

Fig.2 a)Intensity of the shadow graph signal at the leading edge front of the DLTW as a function of time. b) Attractor constructed from the time series $x(t)$ observed at the leading edge front of the DLTW.

3. Summary

We observed a successive transition process of convection patterns in a binary fluid. We studied local dynamics at the fronts in DLTW and found very weak chaotic dynamics. The present system is highly non-linear because of the narrowness of the cell width. The remarkable structure and formation mechanism are originated from such a strong non-linearity. The stability of the fronts is considered to be maintained by the chaotic dynamics at the fronts.

It would be difficult to treat theoretically the situation considered here because of the strong non-linearity. Full numerical calculation of Navier Stokes equations in a two-dimensional space for the strong nonlinear regime has been performed and qualitative feature such as the existence of LTW was reproduced[9]. It needs three dimensional calculations for quantitative comparison with the experiment performed here.

References

[1] M. C. Cross and P. C. Hohenberg, Rev. Mod. Phys. **65** (1993) 851.
[2] J. J. Niemela, G. Ahlers, and D. S. Channel, Phys. Rev. Lett. **64** (1990) 1365.
[3] P. Kolodner, Phys. Rev. Lett. **66** (1991) 1165; Phys. Rev. A **44** (1991) 6448.
[4] W. Barten, M. Lucke, and M. Kamps, Phys. Rev. Lett. **66** (1991) 2621.
[5] H. Riecke, Phys. Rev. Lett. **66** (1992) 301.
[6] P. Kolodner, Phys. Rev. E **48** (1993) R4187.
[7] F. Daviaud, M. Boneti, and M. Dubois, Phys. Rev. A **42** (1990) 3388.
[8] Y. Harada, Y. Masuno, K. Sugihara, K. Nomura, and H. Yahata, in *Pattern Formation in Complex Dissipative systems*, edited by S. Kai (World Scientific, 1992), p.314.
[9] L. Z. Ning, Y. Harada, and H. Yahata, Prog. Theor. Phys. **96** 669.

Analysis of concentration of atmospheric radon daughters as a time series data

Yoichi Tamagawa[1], Tsuguo Nishikawa[1], Takashi Mori[1], Kiyoshi Kudo[1], Ryoku Nakamura[2], and Osamu Yamakawa[3]

1 Dept. of Applied Phys., Faculty of Eng., Fukui Univ. Fukui-shi, Fukui 910, JAPAN
2 Dept. of Phys., Fukui prefectual Univ. Matsuoka, Fukui 910-11, JAPAN
3 Center of Information Science, Fukui Prefectual Univ. Matsuoka, Fukui 910-11, JAPAN

Abstract

A relation of radon (Rn) daughters concentration in the air to meteorological parameters is investigated. The atmospheric radon daughter concentration and the related meteorological parameters are treated as a simple time series data. The power sprctrum of radon daughter concentration and temperature after fourier transfer are fit to 1/f closely. This fact confirms that the common dynamics exist between the variations of the parameters.

Key words: Radon, Meteorological parameters, Fourier transfer, 1/f, Power spectrum

1. Introduction

Radon is a radioactive noble gas spreaded in the atmosphere and plays an important role as the inner radiation source for respiratory organs of human body and the cause of the variation of environmental radiation[1]. The measurement of the radon daughter concentration in the outdoor atmosphere has been carried out in the campus of Fukui University since 1982 .

The relation between the concentration and the related meteorological parameters has already been reported[2]. The correlation coefficients between them, however, are small value and the relation between them are not so tight, because they are considered to be affected by many parameters and have different deley time to each other.

In this paper, the variation of the radon daughter concentration in the atmosphere, the temperature and other parameters are analyzed as the simple series data to investigate the relation between them.

2. What is Radon ?

Radon is one of nuclei of U series. Its half life is 3.8 days with alpha decay. It exhalates from the ground surface to the atmosphere .

As radon has high solbility to the water, it is observed in the groundwater. Recently , the relation between earthquake and radon concentration in the groundwater is reported[3].

The atmospheric radon concentration is not constant, and its time variation is due to the conditions of meteorological parameters (wind velocity, temperature, rain fall, snowfall, e.t.c)[2].

The knowledge of the radon behavior in the atmosphere is very important for the monitoring system around a nuclear facilities because the time variation of the environmental gamma radiation is affected with the radon daughters.

3. Analysis and Results

Fig.1-1 shows a raw data of radon daughter in the air, fig.1-2 shows temperature change. First, the relation between them is investigated directly. This correlation has been showed using a part of these data by selecting the data out of the constant parameter area (i.e. constant wind velocity or constant rain fall) , however, this time, we try not to select data, treat even all data in a month.

First, very weak anti-correlation (fig.2) between a temperature and radon daughter concentration is shown, and no correlation between other parameters is found.

Next, the power spectrum of fourier transfer for this radon daughter concentration is shown in fig.3-1. This is seemed to fit to 1/f plot closely .

And the resut of tempalature data is seemed to fit to the 1/f, too (fig.3-2). This coefficient of this ralation is over 0.8, shows very strong relation .

But, the resuls of other meteorological parameters do not show 1/f fit. These relations are found in all month's data .

This fact seemed that the temperature and radon daughter concentration in the atmosphere are under control of a common dynamics . We think that the inversion layer of air density is one of the sources of these dynamics.

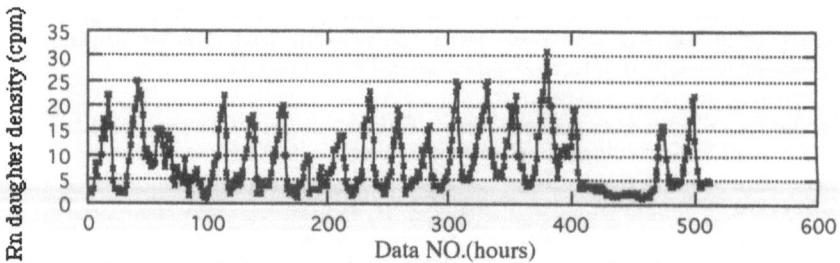

Fig.1-1 Rn concentration data in Augst 1988

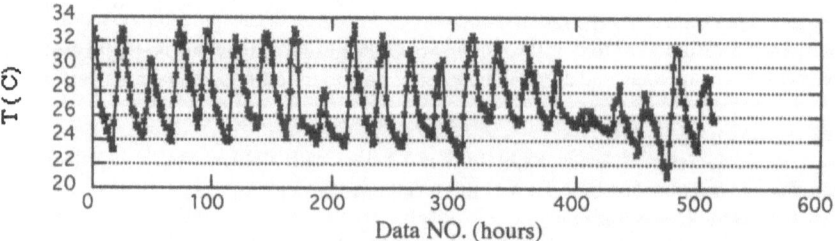

Fig.1-2 Temperature Data in Augst 1988

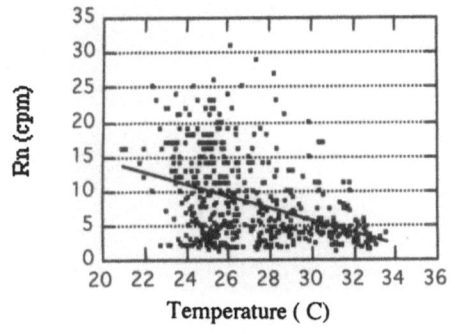

Fig.2 The correlation between
Rn and Temperature

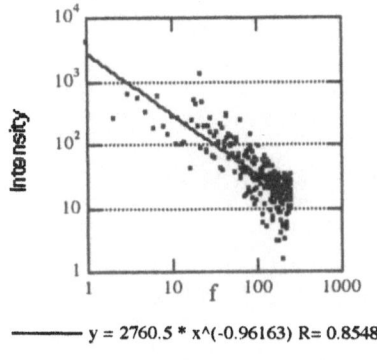

———— y = 2760.5 * x^(-0.96163) R= 0.85485

Fig.3-1
Power Spectrum of Rn
(August 1988)

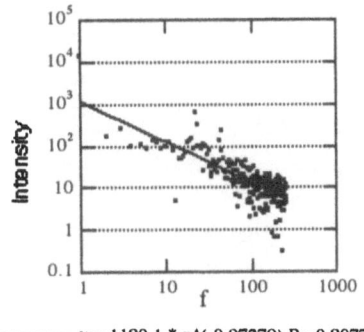

———— y = 1180.1 * x^(-0.97279) R= 0.80777

Fig.3-2
Power Spectrum of Temperature
(August 1988)

4. References

[1] Shigeru Okabe , Introducton to Natural Radon and Its Families, Atmospheric Radon Families
and Environmental Radioactivity , Vol.2, 1990, pp.1-16

[2] Michikuni Simo, Tuguo Nishikawa, General Remarks of Measurement and Analysis,
Atmospheric Radon Families and Environmental Radioactivity , Vol.2, 1990, pp.119-126

[3] G.Igarashi, T.Saeki, N.Takahata, K Sumikawa, S.Tasaka, Y.Sasaki,and Y.Sono, Groundwater
radon anatomy before the Kobe earthquake in Japan, Science, 269, 1995, pp.60-61

Part IV Living Systems

Synchronization in the Brain
- Unification of Information as the Emergence -

Yoko Yamaguchi

Department of Information Sciences, College of Science and Engineering,
Tokyo Denki University, Hatoyama-cho, Hiki-gun, Saitama 350-03, JAPAN

Abstract

Biological complexity derives not only from the variety within the system but also from continuous changes in environments. Synchronization is considered to be the origin of the power by which biological systems describe and control the indefinite order self-organized in themselves. Recognition models based on synchronization of oscillations and based on synchronization of wave propagations are proposed and discussed on their variety and diversity.

Key words: synchronization, oscillation, wave, pattern recognition, speech recognition

1. Complexity in biological systems

When a system contains lots of elements and many interactions among elements, the system has a high degree of freedom in its dynamics. Natural or biological systems are these examples. In order to describe these systems one needs methods how to compress the enourmous amount of information. The description by the order parameter developed in Synergetics is one operatoinal approach to solve the difficulty[1]. If the quality of the order is identified as one type, it can effectively described by the order parameter establishing the description by the external observer.

In biological systems, the description by the oder parameter is not always effective. Further difficulty appears due to the fact that the system faces to continuously changing environments. That is, the boundary condition is not fixed. The indefinite change in the environments makes the order within the system indefinite. Then infinite number of order parameters would be necessary for the external description.

For an organism, the changing orders under the changing environments needs the control by itself so that it should remain to be alive. The biological system needs to control the order of themselves by their own description of the order. The fact that organisms have been alive means that they can describe and hundle their order by themselves.

In order to understand the inteligence of the biological sytem, one has to clarify the internal description within the system. On purpose to understand it, the origin of the power to generate and regulate the order within the system is needed to be considered. We hypotesize that synchronization is the power to unify information instantaneously according to changing environments. In the present paper, we will focus ourselves on the issue how synchronization is generated and controlled in recognition systems.

2. Synchronization in brain dynamics

Synchronization of nonlinear ocsillations in the brain was noted by N. Wiener [2]. He considered the characteristic shape of the spectrum of the human alpha waves as an evidence for synchronization among oscillations of neural activities with various frequencies. An early study on the oscillations in the sensory system is found in the olfactory bulb. W. Freeman [3,4] observed the stimulus dependent activities as synchronization of socillations in the frequency range around 40 Hz by multi-electrode method. He proposes the limit cycle encodes information of the odor . This issue was largely advanced by experimental studies in the visual cortex by Gray, Koenig and Singer in 1989 [5], and Eckhorn et al. in 1988 [6]. They show the unit of oscillations seems to be an orientation column in the visual cortex. Synchronization between a pair of columns is significantly observed when a coherent bar is exposed commonly to their receptive fields. Their results supports the hypothesis that features of sensory information are bound by synchronization encoding a coherent entity as a figure.

From theoretical points of view, the classical neural network architecture that individual entities are encoded in the static state of neural activitiys suffers from the combinatorial expansion and difficulty in the figure/groud separatin. Malsburg (1986) [7,8] showed that temporally correlated neural activities enabled sensory segmentation and separation of the figure from the background. It is extended to the dynamical linking architecture, where the key mechanism of feature linking is fast plasticity of synaptic connections.

The hypothesis of synchronization in the recognition process was theoretically proposed by Shimizu et al. (1985)[9] and developed into a model of pattern recognition with figure/ground separation (Yamaguchi and Shimizu 1994[10,11], Hirakura et al. 1996[12]).

In their hypothesis not only features but also relevancy between sensory information and memory of a symbol are unified by synchronization. Synchronization between the neural oscillators selectively activate units, which further accerelates the synchronization among relevant units. Convergence of linking of the sensory and memory informationto results in the recognition in the complex environment. It takes several cycles of oscillations for synchronization of oscillations to be evaluated to be stable. Thus, recognition is established as a stationary state of synchronization.

3. Feature linking in auditory information processing

In the case of voice recognition, the above hypotehsis could be extended to the speaker-independent vowel recognition in the presence of the noise. Figure 1 shows the vowel recognition model by synchronization by Liu et al. (1995)[13]. A and B centers respectively consist of one-dimensional array of neural oscillators. When the frequency spectrum of the voice is given as input to A center, the local features of the spectrum, formants, evoke local oscillations. The local oscillations are fed to B center, where the memory of vowels are embedded in the connection matrix among neural oscillators in B center. The distributing local features are globally linked according to the memory. The feedback from B to A center amplifys the synchronization among a group of oscillations according to the consistency with the memory. Thus the selective synchronization among formants enables abstraction of formants of a relevant vowel in the presence of unbiguity of the spectrum or noises.

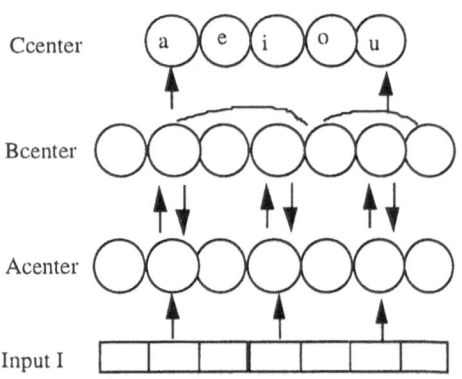

Fig.1 Model of vowel recognition. A and B centers consists of the neural oscillator units. A center has latral inhibitions for feature extraction. B center has long range connections for global linking.

However, the binding by synchronization

cannot be extended straightforward to the speech recognition with vowels and consonants. It is because features of consonants are essentially transient. If the consonants are processed by the synchronization hypothesis, features of consonants cannot be evaluated by the time scale of the vowels.

Does it need another network for the linking in another time scale? The democposition of the networks into different ones for different features makes another difficulty for unification of the networks. In the following section we elucidate the linking of the transient features by studying dynamics in the auditory cortex.

4. A model of wave propagation in the auditory cortex

Our present model for the processing the transient features is inspired by the experimental observation of the guinea pig auditory cortex by optical recording (Taniguchi et al. , 1992[14]). The experimental result is shematically shown in Fig. 2. When a pure tone with constant frequency (CF) is given as a stimulus to the auditory system , an active spot in the auditory cortex start at one end propagating along an isofrequency band. On the other hand, the stimulus of frequency modulation (FM) brings on an active spot propagating from one end to the other across isofrequency bands. These observations show that the features of the auditory information CF/FM is represented by collective spatio-temporal patterns of the neural network. The propagation trace of the active spot does draw a kind of sonogram on the plane of the auditory cortex, either of field A or DC.

We propose a simple neural network model of the wave propagation in the auditory cortex. Figure 3 show the schematic illustration of the model. It consists of a two dimensional array of neural oscillators. One axis of the two dimension is assumed to have the tonotopy as topographic mapping of input with frequency selectivity. The other axis shall be called propagation axis for the sake of convention. Each neural oscillator has neighboring excitatory connections along

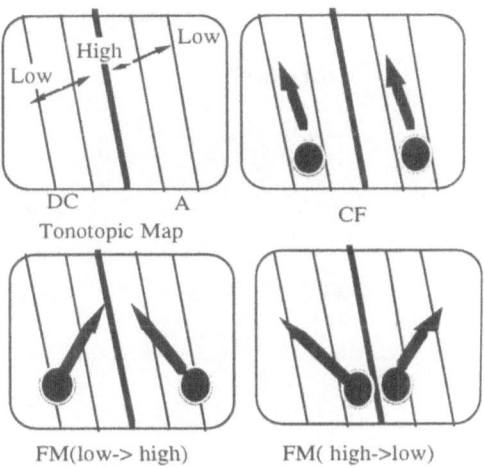

Fig. 2 Observation of Spatio-Temporal Pattern of Frequencty Representation by Optical Recording in the Auditory Cortex of Guinea Pigs [14].

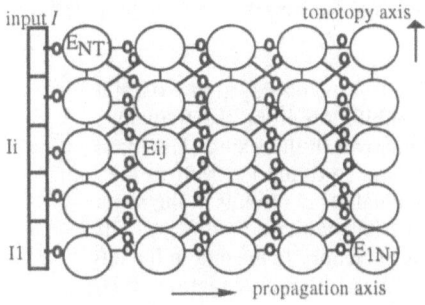

a two-dimensional neural network

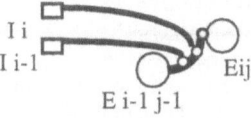

Tonotopic input modifying the neighboring connection

Fig. 3 Structure of the model of wave propagation

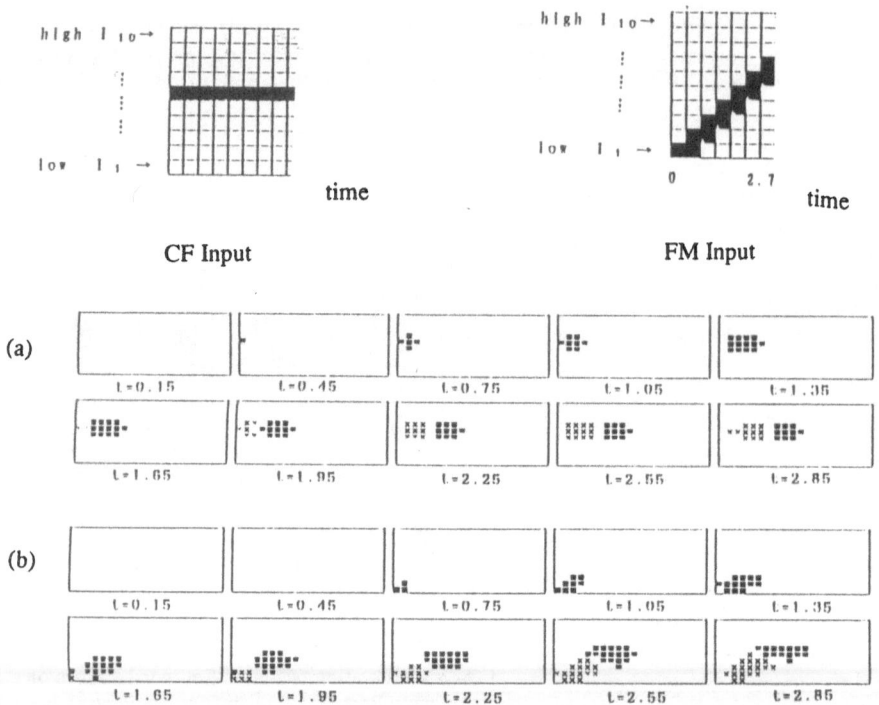

Fig. 4 Results of computer experiments. (a) and (b) show theTime evolution of neural unit activities with CF and FM input respectively. Filled sqares show active units. Cross simbols show units are suppressed(or hyperpolarized).

the direction of propagation axis and lateral inhibition along the frequency axis as shown in the figure. The input of the tone stimulus I is sent to one-dimensional array of units at the same tonotopy axis with no time delay. The input is excitatory to the unit in the left hand side of the propagation axis, while other units have no direct input but presynaptic modification by the input I in the neighboring excitatory connections, as shown in the figure.

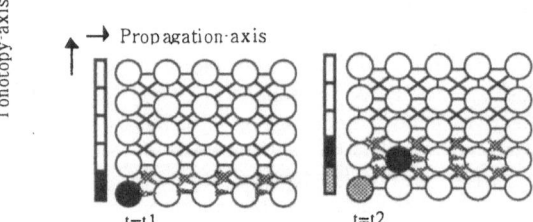

Fig. 5 Schematic illustation of wave propagation with FM input. Filled circles and gray arrows denote active units and potentiated synapses respectively.

The above neural network model is mathematically formulated as a set of ordinary differential equations [15,16]. The values of parameters are chosen so that the neural units has the excitability to give the wave propagation in the presence of input I and synaptic modification.

The computer experiments by using the above equations were carried out by Runge-Kutta-Gill method on PC98 AP. Figure 4 show the results with input stimuli CF and FM. It is seen that the

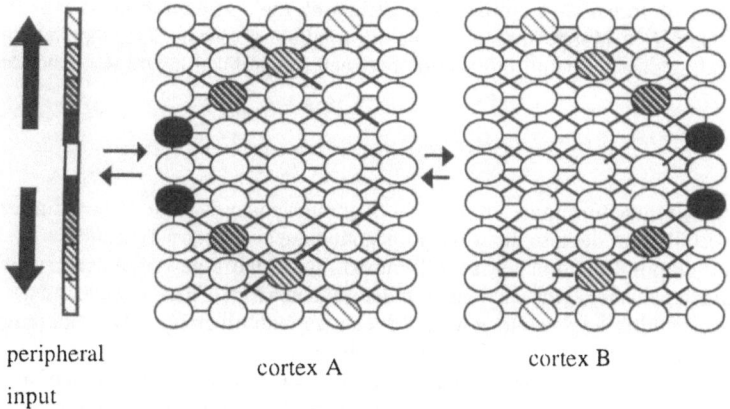

peripheral
input

cortex A

cortex B

Fig. 6 Information processing by spatio-temporal Pattern of Neural Activities.
Feature abstraction is processed by generation and extinction of wave propagation in
cortex A. Feature binding is controlled by entrainment of wave propagations within the
cortex and between the two cortices.

results of the optical recording have been qualitatively reconstructed.

The propagation in this model is essentially anisotropic and localized. Localization of the activities on the frequency axis encodes instantaneous information of the stimulus I . On the propagation axis, it depends on the preceding spatial pattern. That is, the network represents the instantaneous and context dependent information on each axis, as illustrated in Fig. 5.

Further computer experiments show that the spatio-temporal pattern generated on the network plane is characterized by the nonlinearity of the dynamics of the network. We note that those characteristics have some correlation with the transient characteristics of consonants.

First, the life span of the wave propagation is sensitive to some characteristic feature of speech. When CF is followed by immediate FM with no time delay and no frequency gap, the wave propagation continues only when the modulation rate is less than a critical value. The rapid modulation above the threshold makes extinction of the propagation. It derives from threshold phenomena of the excitation of the neural unit. The extinction of propagation means the loss or resetting of the context in the network. In this situation next input stimulus generates new active spot at the end of the propagation axis. It means that resetting of the context depends on the FM rate. The resetting is also regulated by the length of the silent interval between tone stimuli.

Thus, Transient FM and silent interval that are characteristics of consonants are found dominant factors of resetting of context. In the case of speech of vowel-consonant-vowel (VCV), its articulation to V and CV shall be possible by the segmentation of the wave propagation for a formant into two pieces caused by the transient features of C. Though V or CV consists of three or four formants, only one formant is considered in the above.

Binding of the plural formants is now an issue of synchronization of wave propagation. Results of computer experiments show that a few pieces of waves are pulled together on the propagation axis under some condition. Such a type of synchronization naturally originates from the stability of the plane wave, that is known in an isotropic excitable medium. The pieces of wave cooperates through interactions on the network giving rise to "the coherent contextual information".

When this model is applied to recognition of a syllable, further control for segmentation and synchronization of wave propagation should be necessary according to the memory of speech. Then the hierarchical structure shown above is necessary to be considered. The wave propagations encode elementary information of transient features. The ability of the segmentation of waves and synchronization of propagation among waves unify the information of speech. Thus the recognition by feature binding through circulation of information between hierarchical layers was extended to speech recognition.

5. Conclusion

We discussed the synchronization in the brain as the key mechanism for the feature binding in recognition. Not only oscillations but also the wave propagation are elementary dynamics in the self-organization of spatio temporal pattern. As is well known, these two types of dynamics are not exclusive but alternatively generated with each other in the same network. These nonlinear dynamics have the stability and flexibility to generate coherent dynamics with diversity, which supports the biological systems to unify information with flexibility and diversity.

It is important to clarify how the synchronization is controlled by the system. Elucidation of the way of self-control of synchronization would enlighten how semantic information is stored and how emergent information is created in biological systems.

Acknowledgement

The author would like to express appreciation in the above study of auditory system, to Prof. I. Taniguchi for valuable contribution, to Prof. R. Suzuki for discussion and encouragement, and to Mr. Y. Emae and Mr. T. Hasebe for their cooperation in the computer experiments.

References

[1] Haken, H.ed. (1985) Complex Systems- Operational Approaches in Neurobiology, Physics, and Computers, Springer.

[2] Wiener, N.(1958) Nonlinear Problems in Random Theory. MIT Press, Cambridge, MA.

[3] Freeman, W. J. (1975) Mass action in the nervous system. Academic Press, New York.

[4] Skarda, C.A. and Freeman, W. J. (1987) How brains make chaos in order to make sense of the world, Behavioral adn Brain Sciences, 10, 161-195.

[5] Gray, C. M., Koenig, P., Engel, K. P. & Singer, W.(1989). Oscillatory responses in cat visual cortex exhibit inter-columnar synchronization which reflects global stimuli properties. Nature, 338, 334-337.

[6] Eckhorn, R., Bauer, R, M., Jordan, W., Brosch, M., Kruse, W., Munk, M. & Reitboeck et al.(1988). Coherent Oscillations: A Mechanism of Feature Linking in the Visual Cortex? Biological Cybernetics, 60, 121-130.

[7] von der Malsburg, C. (1986) Am I Thinking Assemblies? In G. Palm and A. Aertsen, (Eds.), Brain Theory, (161-176). Berlin, CA: Springer.

[8] von der Malsburg, C. & Schneider, W. (1986) A neural cocktail-party processor. Biological Cybernetics, 54, 29-40.

[9] Shimizu, H., Y. Yamaguchi, I. Tsuda and M. Yano (1985) "Pattern Recognition Based onHolonic Information Dynamics towards Synergetic Computers", in Complex Systems-Operational Approaches in Neurobiology, Physics, and Computers. ed. H. Haken, Springer, 225-239.

[10] Yamaguchi, Y. and H. Shimizu (1994) Pattern recognition with Figure-Ground Separation by generation of coherent oscilaltions, Neural Networks,7(1) pp.49-6.

[11] Yamaguchi, Y. and H. Shimizu (1994) "Self-organization of Information in the Brain", in Physics of Living State., T. Musha and Y.Sawada eds. Ohm-Sya, Tokyo, 253-286.

[12] Hirakura,Y.,Y. Yamaguchi , H. Shimizu and S. Nagai (1996) Dynamic Linking among Neural Oscillators leads to Flexible Pattern Recognition with Figure Ground Separation, Neural Networks , 9(2) pp.189-209.

[13] Liu, F. , Y. Yamaguchi and H. Shimizu (1994) Flexible vowel recognition by the generation of dynamic coherence in oscillator neural networks: speaker-independent vowel recognition, Biol. Cybern.71:105-114.

[14] Taniguchi, Ikuo, Junsei Horikawa, Toshio Moriyama and Masahiro Nasu (1992) Spatio-temporal patternof frequency represen-tation in the auditory cortex of guinea pigs, Neuroscience Letters, 146 37-40.

[15] Yamaguchi, Y. and I. Taniguchi, Neural Activities as Wave Propagation in a Neural Network Model of the Auditory Cortex, (1996) Technical Report of IEICE, NC95-141, 197-204.

[16] Yamaguchi, Y., Y. Emae, K. Machida, Y. Marumo, T. Hitomi and I. Taniguchi, Neural Activities as Wave Propagationand Feature Abstraction in the Auditory Cortex, (1996) Trans tech. Com. Psycho. Physio. Acoust. H-96-33.

The developing *Xenopus* embryo as a complex system: Maternal and zygotic contribution of gene products in nucleo-cytoplasmic and cell-to-cell interactions

Koichiro Shiokawa, Hiroshi Fukamachi, Chie Koga, Naoki Adati, Miyuki Amano, Jun Shinga, Mikihito Shibata, and Yoichi Yamada

Laboratory of Molecular Embryology, Department of Biological Sciences, Graduate School of Science, University of Tokyo, Bunkyo-ku, Tokyo 113, JAPAN

Abstract

The developing animal embryo constitutes a complex system in which various nucleo-cytoplasmic (N-C) and cell-to-cell (C-C) interactions take place. In that sense, it is possible to define early embryogenesis as a function of these interactions, as for instance is expressed by a formula "Development = f (N-C, C-C)". We present here our recent studies on temporal and spatial control of the expression of genes in zygotic nucleus and of genes exogenously introduced in *Xenopus* embryos. For zygotic gene expression, our studies revealed that the syntheses of mRNA, tRNA and rRNA are initiated at the cleavage stage, the stage of midblastula transition (MBT) and late blastula stage, respectively. For exogenously-injected genes, we summarize their expression pattern which is controlled by the promoter they carry in addition to the cytological effects of the injection. We also briefly present our recent results obtained with embryos which had been injected with *in vitro*-transcribed mRNAs.

Key words: *Xenopus* embryos, maternal RNA, midblastula transition, zygotic transcription, DNA and RNA microinjection,

1. Introduction

Over thirty years ago, Brown and Littna [1] extracted RNAs by phenol method from frog embryos and started to handle undegraded, as opposed to alkaline-hydrolyzed PCA or TCA-extracted, RNAs from developing frog embryos. The impermeable surface coat covering the embryo prevents the uptake of RNA precursors administrated to their culture medium. Brown and Littna [1] overcame the difficulty by injecting ^{32}P-orthophosphate into gravid *Xenopus laevis* females, and followed accumulation of the labeled RNA in the course of the development. Shiokawa and Yamana [2] did so by labeling embryos with $^{14}CO_2$ gas or by using dissociated embryonic cells which efficiently uptake and incorporate specific RNA precursors from the culture medium. The pattern of RNA synthesis in *Xenopus* embryos was studied in these ways, and the mechanism of gene expression from zygotic nucleus (nuclear DNA) in early amphibian embryos started to be explored.

In the earliest phase of these studies, it was found that pattern of RNA synthesis changes from blastula-type (high 4S RNA or tRNA synthesis) to neurula-type (high rRNA synthesis) in whole embryos as well as in dissociated embryonic cells. Subsequent studies have refined our view of the developmental changes in gene expression in *Xenopus laevis* using more sophisticated methods, including cloning and microinjection of the single-isolated genes and their products (mRNAs and proteins). Thus, it was found that RNA synthetic activity is enhanced (at least on a per embryo basis) greatly at the midblastula stage (just after 12 cleavages) being accompanied with two major changes in cellular activities, acquisition of cell motility and lengthening (or slowing) of cell cycle, giving rise to G1 phase. The phenomenon was termed midblastula transition (MBT) [3] and is now widely known, although the validity of this concept has been discussed as with whether or not it is acceptable in its original form [4] .

In this article, we summarize our studies on the temporal and spatial changes in RNA synthetic pattern based on zygotic nucleus and then proceed to more recent data on the fate during *Xenopus* embryogenesis of exogenously-injected DNAs as well of exogenously-injected mRNAs like activin receptor and S-adenosylmethionine decarboxylase (SAMDC). Thus, we will try to emphasize the importance of N-C and C-C interactions together with that of the mechanism of switch from maternal to zygotic gene expression.

2. Contribution of maternal and zygotic mRNAs to early embryonic development

All multicellular animals arise from single cells called fertilized eggs. Fertilizable *Xenopus* eggs are formed from full-grown oocytes after meiotic division called oocyte maturation. During the maturation which lasts for about 10 hr, nuclear activity is relatively small. The cytoplasm of *Xenopus* oocytes as well as eggs contain a large amount of maternally-formed mRNA, which comprises approximately 80 ng (the content of total RNA is ca. 4000 ng per egg [1, 2]) [5] . These maternally-inherited mRNAs start to be translated from the time of the onset of oocyte maturation and it is believed that events in the earliest phase of the development are supported by the functioning of these translated materials, which include proteins of both house-keeping and morphogenetic nature. In Fig. 1, we present the data which shows occurrence of maternally-inherited mRNA for type C aldolase [6] , an enzyme necessary for glycolysis that gives rise to the energy source. Also, Fig. 2 shows a relatively uniform distribution within the early *Xenopus* embryo of another maternal mRNA called activin receptor (type IIA and IIB), together with the localized expression pattern in later stage of development of a zygotically-expressed mRNA called follistatin [7] which is needed for morphogenesis, probably mresoderm formation, which starts at and after the blastula stage. Figure 3 shows changes in a subset of maternal as well as zygotic transcripts studied by a more recent technique called fluorescent differential display [8] , which clearly gives us an idea that many RNA molecules which are present in the fertilized eggs are consumed and new transcritps appear newly in the course of the development. It is believed that embryos may not undergo normal development, unless these and other maternal mRNAs are properly supplied and new zygotic transcription commences in due time. These data especially those in Fig. 3 exemplify the complexity seen in the changing population of mRNAs in early developing embryos.

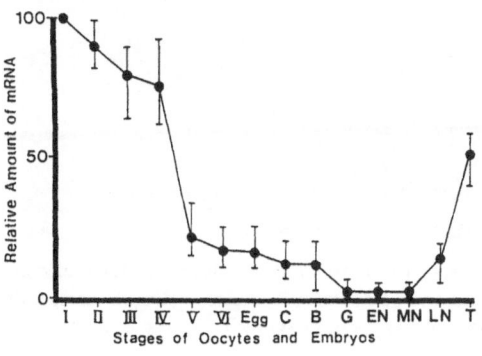

Fig. 1 Changes in the level of *Xenopus* type C aldolase mRNA during oogenesis and early embryogenesis, analyzed by Northern blotting (normalized to that of stage I oocytes). From Atsuchi et al. [6] .

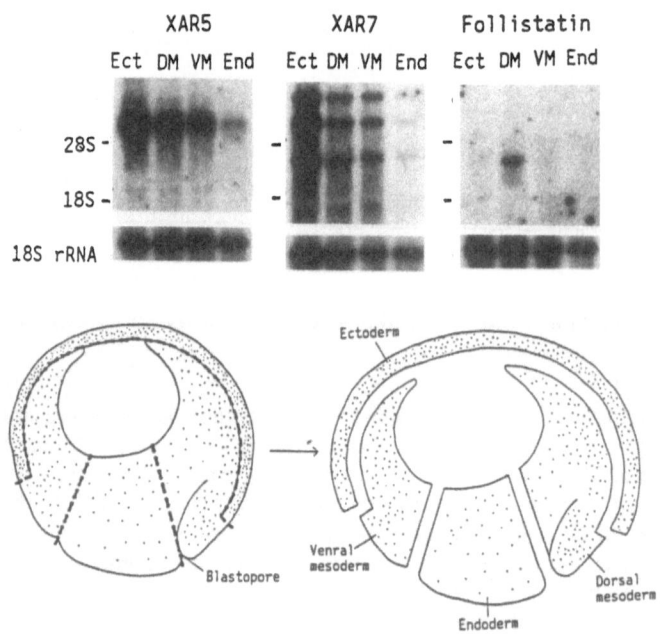

Fig. 2 Spatial distribution of mRNAs of *Xenopus* activin receptor type IIB (XAR5), type IIA (XAR7) and follistatin at the gastrula stage. Strong signal for activin receptor mRNAs was found in ectoderm (Ect), followed by dorsal mesoderm (DM), ventral mesoderm (VM) and endoderm (End). Follistatin mRNA was found only in dorsal mesoderm. From Koga et al. [7].

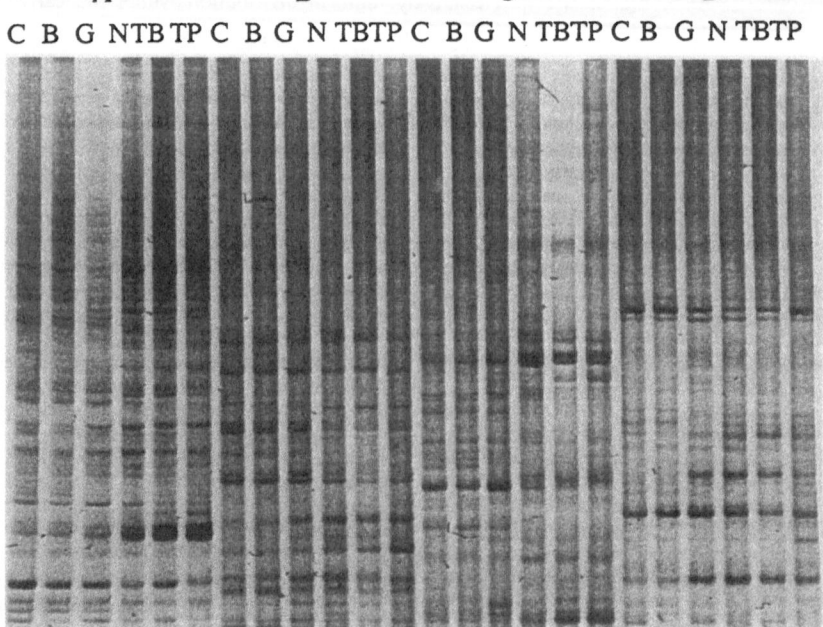

Fig. 3 Fluorescent differential display of RNAs from *Xenopus* embryos at (C) cleavage, (B) blastula, (G) gastrula, (N) neurula, (TB) tailbud and (TB) tadpole stages. Results using different arbitrary primers were shown in A, B, C and D. From Adati et al. [8].

3. Changing transcriptional activities from zygotic nucleus mediated by three RNA polymerases in early embryos

RNA synthesis is mediated by three RNA polymerases I, II and III, which respectively transcribe rRNA, mRNA and tRNA.   To know activity of these enzymes, it is important to clearly define the RNA species newly transcribed in early embryogenesis.   In our previous studies, we observed methylation of mRNA CAP in cleavage stage embryos, which predicted the onset of mRNA synthesis in these very early embryos [9].   We labeled here RNA with (methyl-^{3}H)-methionine in embryos which were dissociated within the vitelline membrane by cultivation in an isotonic phosphate buffer solution.   Furthermore, we detected the syntheses of both heterogeneous mRNA-like RNAs and a relatively small amount of low-molecular-weight RNAs in the dissociated cleavage-stage embryos by labeling them with ^{3}H-uridine.   This latter finding is compatible with those of Kimelman et al.[10] who analyzed RNA synthesis by microinjecting ^{32}P-GTP into fertilized eggs.   From these experiments, we reached the conclusion that pre-MBT-stage embryos synthesize heterogenous mRNA-like RNA (already at the 64-cell stage), and in addition, a small amount of small-molecular-weight RNAs during the pre-MBT stage [11].   The kinetics of the accumulation of the RNAs was roughly proportional to the cell number before and after the MBT.   However, as for 4S RNA, the rate of its synthesis was found to increase approximately by 100-fold on a per-cell basis at the MBT stage, although the increase in the activity per cell of the synthesis of other RNAs, such as heterogenous mRNA-like RNA and small-molecular-weight RNAs (5S RNA and snRNAs), was only 2- to 3-fold.

The distinct peaks of 18S and 28S rRNA labeling were initially detected at the gastrula stage [1, 2].   However, since the rate of mRNA synthesis per cell is extremely high at the blastula stage (10-fold or more than that in gastrula and neurula stages) [5], it was suspected that peaks of rRNA synthesized at this stage may be dismissed by the presence of a large and broad peaks of mRNA-like RNAs.   Therefore, to specify the stage when rRNA synthesis initiates more precisely, dissociated embryonic cells were labeled again with (methyl-^{3}H)-methionine and 2'-O-methylation of high-molecular-weight RNAs was examined [9], since this occurs in rRNA but not in mRNA.   When labeled high-molecular-weight RNAs were analyzed on DEAE-Sephadex columns after complete digestion with RNases A and T2, RNA from cleavage stage embryos (labeled for 3 h) gave rise to methylation peaks of charge -5 (mRNA CAP), but not charge -3 (NpmNp) and -4 (NpmNpmNp)(both rRNA-derived), nucleotides.   When early blastula cells were labeled for 4 h until the late blastula stage, however, a very large peak of rRNA-specific charge -3 and a small peak of -4 components were obtained.   By similar experiments in which we divided the 4 h of blastula stage into two 2 h-periods, we found that embryos start to synthesize rRNA not in the former half but in the latter half of the blastula stage (late blastula stage).   The timing of the first appearance of definitive nucleoli studied cytologically [12] coincided with this result.   Since 2'-O-methylation occurs on newly transcribed rRNA, we concluded that rRNA synthesis is initiated from the late blastula stage.

Based on these results, we concluded that *Xenopus* early embryogenesis consists of at least three different phases with respect to the pattern of RNA synthesis (Fig. 4) [13].   The first is the pre-MBT stage, which is characterized by a relatively high activity (on a per-cell but not per-embryo basis) of mRNA-like RNA synthesis and also by a relatively low activity of low-molecular-weight RNA synthesis.   The second is the MBT stage, which is characterized by a great activation of the synthesis of 4S RNA.   The third is the post-MBT stage, which is characterized by a gradual activation (per embryo) of the synthesis of rRNA.   Thus, RNA synthetic patterns at pre-MBT, MBT and post-MBT phases are characterized by predominant activity of RNA polymerases II, III and I, respectively.

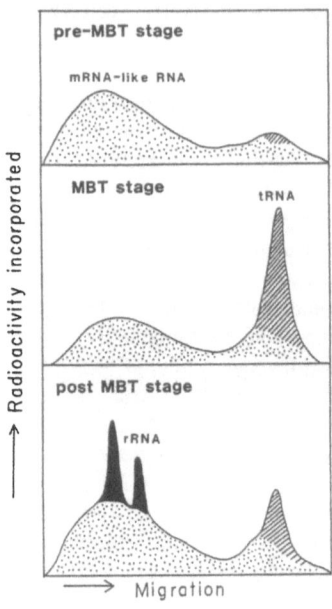

Fig. 4 Three characteristic profiles of RNA synthetic patterns in *Xenopus* early embryos. Dotted, shaded and black areas represent products of RNA polymerase II, III and I, respectively. From Shiokawa et al. [13].

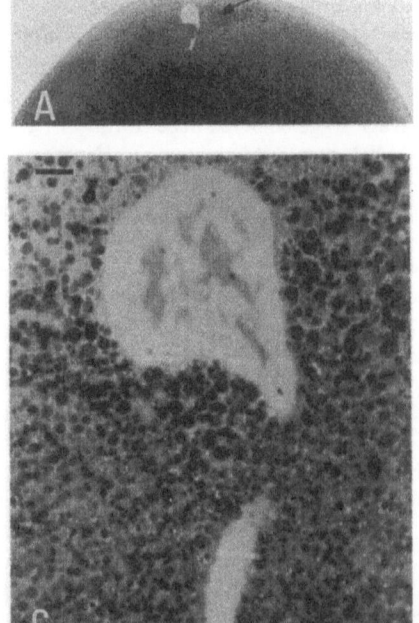

Fig. 5 Light micrographs of nucleus-like structures formed by microinjecting (A) 24 ng/egg of lambda DNA into fertilized *Xenopus* eggs. (C) is a high power magnification of (A). Bars; (A) 100 μ m and (C) 10 μ m. From Shiokawa et al. [14].

4. Expression of exogenously-injected genes

Unlike mouse eggs, *Xenopus* eggs are not transparent.   Therefore, it is not possible to inject DNA (or genes) into zygotic nucleus.   This means that when we inject exogenous DNA (or genes) into the *Xenopus* fertilized egg, the DNA is injected into the cytoplasm.   Unlike the oocyte cytoplasm, egg cytoplasm does not degrade the injected DNA.   Furthermore, injected DNAs, both linear (eg. lambda DNA) and circular (eg. pBR322), are associated with maternal histone and assembled in the form of nucleus-like structure (Fig. 5) surrounded by the normal-looking nuclear envelope which is provided with numerous normal-looking nuclear pore complexes [14].   Autoradiographic tracing revealed that the injected DNA in such nucleus-like structures is partitioned in about 70% of the descendant cells when the dosage of the DNA was not high (less than 2ng/egg).

When we injected various circular plasmids that carried bacterial chloramphenicol acetyltransferase (CAT) gene at 1ng/egg (ca. 10^7 copies/egg), the injected genes were found to be expressed.   In such injection studies, the extent of the expression depended on the strength of the promoter included in the plasmid [15].   In the experiment of Etkin and Balcells [16]who injected 1 ng/egg of pSV2CAT that contained the SV40 early promoter, it was reported that CAT gene expression takes place only after 12th cleavage (namely at the MBT).   However, when we repeated the experiment after increasing the amount of pSV2CAT by 10-fold or after increasing the number of embryos per sample by 10-fold (100 embryos/sample) without increasing the dosage, the CAT enzyme activity was detected not only at and after MBT but also at 3 h (early cleavage) and 4 h (late cleavage stage) after fertilization (Fig. 6 left) [17] .   It was found here that the increase in the expressed CAT enzyme activity roughly paralleled the increase in the number of cells per embryo.

For the sake of comparison, we also tested the expression of a CAT gene #254, which contained the promoter of *Xenopus* α-actin gene, at doses of 1 ng/egg and 10 ng/egg.   By these injection experiments, we found that the injected CAT gene with the actin promoter is always expressed at and after the neurula stage (Fig. 6 right) [17] , just like its endogenous counterpart [18] .   Temporally regulated expression was also observed with CAT gene with *Xenopus* HGF promoter [19] .   These results indicate that under our conditions the expression of the gene which has the promoter of a temporally controlled *Xenopus* gene is expressed at the correct timing during development.

Finally, as another point of interest, we tested if the injected DNAs are replicated within *Xenopus* embryos.   The results showed that while the circular DNAs poorly replicate, linear DNAs replicate quite actively and form concatemers of molecular size which is much longer than the injected DNA [15].   Furthermore, we noticed that the timing of the expression changes (becomes earlier) when CAT gene with developmentally-controlled promoter was injected after being linearized [20] .

5. Expression of exogenously-injected mRNA

Along with the experiments to inject purified DNAs, we also carried out injection of the specific mRNA which had been transcribed *in vitro* from the cDNA clones.   Among these, we obtained relatively clear-cut morphological effect, when we injected the mRNA of type IIA activin receptor cloned in our laboratory [21] .   As shown in Fig. 7A, overexpression of this mRNA, which is provided as a maternal mRNA at a relatively high level [7, 21] , induced the formation of the secondary body axis and/or tail-like structure, when the injection was carried out into the ventral blastomeres at the cleavage stage.   This mRNA codes for a transmembrane protein, which activates the intracellular signal transduction by its cytoplasmic serine-threonine kinase domain, as has been shown with type IIB in other laboratory [22] .   When we constructed type IIA mRNA lacking the kinase domain and injected it into the presumptive dorsal side of the fertilized eggs, embryos without

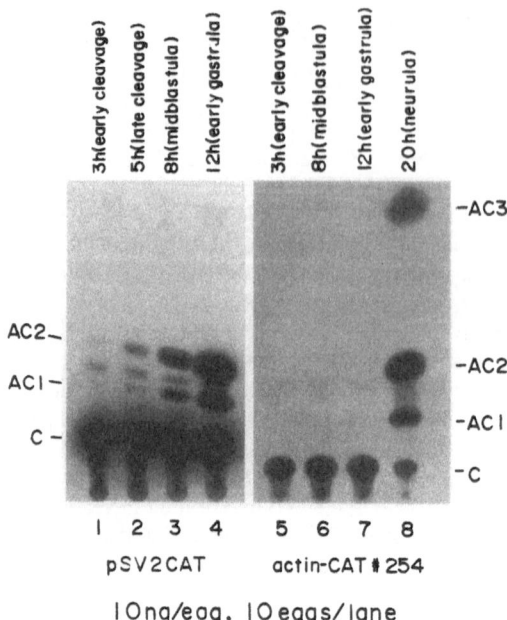

Fig. 6 CAT gene expression by microinjecting pSV2CAT (left) or actin-CAT#254 (right) genes into fertilized *Xenopus* eggs. CAT enzyme activity could be detected from early cleavage stage (3h) onwards when pSV2CAT was injected, whereas it was found only at and after neurula stage when actin-CAT#254 was injected. From Shiokawa et al. [17].

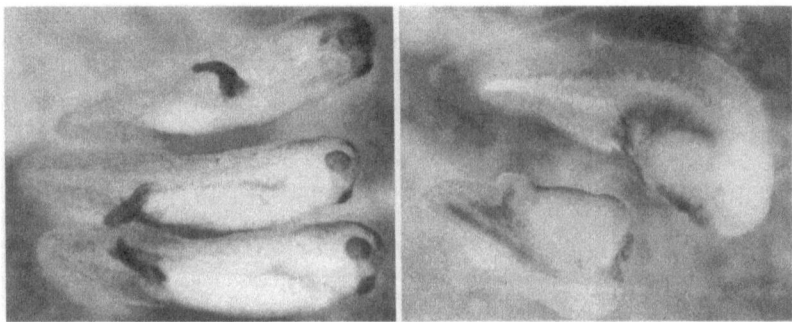

Fig. 7 *Xenopus* embryos injected with mRNA of type IIA activin receptor. Left, wild type mRNA. Right, truncated mRNA, lacking the serine-threonine kinase domain. M. Amano et al. (unpublished data).

head part were obtained (Fig. 7B). These results suggest the distortion of the normally occurring cell-to-cell interaction which is necessary for the formation of head part by the truncated receptors [cf. 22]. Thus, the results in Fig. 7 provide examples which indicate the importance of cell-to-cell interaction in the morphogenetic activities during the development.

In relation to the function of exogenously-injected mRNAs, we recently cloned SAMDC mRNA [23] and injected its mRNA into fertilized eggs, with the result that the injected embryos arrested its development specifically at the gastrula stage. Polyamines may give the circumstantial conditions which permit various molecules to behave properly.   Therefore, the elucidation of the mechanism of the arrest may be expected to help understand the mechanism of development and its complexity.

6. Conclusion

The developing *Xenopus* embryo is a complex system in which maternal gene products are substituted by zygotic gene products.   Also, the developing embryo constitutes a complex system in which both nucleo-cytoplasmic and cell-to-cell interactions take place.

7. References

[1] D. D. Brown, E. Littna, RNA synthesis during the development of *Xenopus laevis*, the African Clawed toad, *Journal of Molecular Biology*, Vol. 8, 1964, pp. 669-687.

[2] K. Shiokawa, K. Yamana, Pattern of RNA synthesis in isolated cells of *Xenopus laevis* embryos.   *Developmental Biology*, Vol. 16, 1967, pp. 368-388.

[3] J. Newport, M. Kirschner, A major developmental transition in early *Xenopus* embryos, I. Characterization and timing of cellular changes at the midblastula stage, *Cell*, Vol. 30, 1982, pp. 675-686.

[4] G. K. Yasuda, G. Schubiger, Temporal regulation in the early embryo: is MBT too good to be true, *Trends in Genetics*, Vol. 8, 1992, pp. 124-127.

[5] N. Sagata, T. Nakahashi, K. Shiokawa, K, Yamana, Poly(A)-containing RNA synthesis in *Xenopus laevis*, *Cell Structure and Function*, Vol. 3, 1978, pp. 71-78.

[6] Y. Atsuchi, K. Yamana, H. Yatsuki, K. Hori, S. Ueda, K. Shiokawa, Cloning of a brain-type aldolase cDNA and changes in its mRNA level duriing oogenesis and early embryogenesis in *Xenopus laevis*.   *Biochimica et Biophysica Acta*, Vol. 1218, 1994, pp. 153-157.

[7] C. Koga, K. Tashiro, K. Shiokawa, Different spatial distribution of mRNAs for activin receptors (type IIA and IIB) and follistatin in developing embryos of *Xenopus laevis*. *Roux's Archives of Developmental Biology*, Vol. 204, 1995, pp. 172-179.

[8] N. Adati, T. Ito, C. Koga, K. Kito, Y. Sakai, K. Shiokawa, Differential display analysis of gene expression in developing embryos of *Xenopus laevis*, *Biochimica Biophysica Acta*, Vol. 1262, 1995, pp. 43-51.

[9] K. Shiokawa, Y. Misumi, K. Yamana, Demonstration of rRNA synthesis in pre-gastrular embryos of *Xenopus laevis*, *Development, Growth and Differentiation*, Vol. 23, 1981, pp. 579-587.

[10] D. Kimelman, M. Kirshner, T. Scherson, The events of the midblastula transition in *Xenopus* are regulated by changes in the cell cycle, *Cell*, Vol. 48, 1987, pp. 399-407.

[11] K. Shiokawa, Gene expression from endogenous and exogenouly-introduced DNAs in early embryogenesis of *Xenopus laevis*, *Development, Growth and Differentiation*, Vol. 33, 1991, pp. 1-8.

[12] T. Nakahashi, K. Yamana, Biochemical and cytological examination of the initiation of ribosomal RNA synthesis during gastrulation of *Xenopus laevis*, *Development, Growth and Differentiation*, Vol. 18, 1976, pp. 329-339.

[13] K. Shiokawa, R. Kurashima, J. Singa, Temporal control of gene expression from endogenous and exogenously-introduced DNAs in early embryogenesis of *Xenopus laevis*, *International Journal of Developmental Biology*, Vol. 38, 1994, pp. 249-255.

[14] K. Shiokawa, M. Yoshida, H. Fukamachi, Y. Fu, K. Tashiro, M. Sameshima, Cytological studies of large nucleus-like structures formed by exogenously-injected linear and circular DNAs in fertilized eggs of *Xenopus laevis*, *Development, Growth and Differentiation*, Vol. 34,

1992, pp. 79-80.

[15] Y. Fu, K. Hosokawa, K. Shiokawa, Expression of circular and linearlized bacterial chloramphenicol acetyltransferase gene with or without viral promoters after injection into fertilized eggs, unfertilized eggs and oocytes of *Xenopus laevis*, *Roux's Archives of Developmental Biology*, Vol. 198, 1989, pp. 148-156.

[16] L. D. Etkin, S. Balcells, Transformed *Xenopus* embryo as a transient expression system to analyze gene expression at the midblastula transition, *Developmental Biology*, Vol. 108, 1985, pp. 173-178.

[17] K. Shiokawa, K. Yamana, Y, Fu, Y, Atsuchi, K. Hosokawa, Expression of exogenously-introduced bacterial chloramphenicol acetyltransferase gene in *Xenopus laevis* embryos before the midblastula transition, *Roux's Archives of Developmental Biology*, Vol. 198, 1990, pp. 322-329.

[18] T. J. Mohun, N. Garrett, J. B. Gurdon, Upstream sequences required for tissue-specific activation of the cardiac actin gene in *Xenopus laevis* embryos, *EMBO Journal*, Vol. 5, 1986, pp. 3185-3193.

[19] H. Nakamura, T. Tashiro, K. Shiokawa, Isolation of *Xenopus* HGF gene promoter and its functional analysis in embryos and animal caps. *Roux's Archives of Developmental Biology*, Vol. 205, 1996, pp. 300-310.

[20] K. Shiokawa, Y. Fu, K. Hosokawa, K. Yamana, Temporally uncontrolled expression of linearized plasmid DNA which carries bacterial chloramphenicol acetyltransferase gene with *Xenopus* cardiac alpha-actin promoter after injection into *Xenopus* fertilized eggs. *Roux's Archives of Developmental Biology*, Vol. 199, 1990, pp. 174-180.

[21] M. Kondo, K. Tashiro, G. Fujii, M. Asano, R. Miyoshi, R. Yamada, M. Muramatsu, K. Shiokawa, Activin receptor mRNA is expressed early in *Xenopus* embryogenesis and the level of the expression affects the body axis formation. *Biochemical and Biophysical Research Communications*, Vol. 181, 1991, pp. 684-690.

[22] A. Hemmati-Brivanlou, D. A. Melton, A truncated activin receptor inhibits mesoderm induction and formation of axial structures in *Xenopus* embryos, *Nature*, Vol. 359, 1992, pp. 607-614.

[23] J. Shinga, K. Kashiwagi, K. Tashiro, K. Igarashi, K. Shiokawa, Maternal and zygotic expression of mRNA for S-adenosylmethionine decarboxylase and its relevance to the unique polyamine composition in *Xenopus* oocytes and embryos, *Biochimica Biophysica Acta*, Vol. 1308, 1996, pp. 31-40.

Universal feature of long-range correlations in DNA sequence : The intron-structure in a single gene

S. Tanda and T. Teramoto

Department of Applied Physics, Hokkaido University, Sapporo 060, Japan

Abstract

We investigated the power spectrum of *the intron-structure in a single gene* over 70 genes in 9 biological classifications. Our results provided clear evidence that *the intron-structure* universally show 1/f spectrum in all classifications. To confirm this result, we showed that the long-range correlation is not effected where concerned with more than 2 genes. Universal of long-range correlations are connected with the origin of intron, i.e., the intron-late view. We suggest that the intron insertion system in a single gene is the origin of long-range correlations in DNA sequence, and that there is a possibility of self organized criticality in molecular evolution.

Key words: DNA, 1/f noise, dynamic scaling, long-range-correlation, evolution

1. Introduction

The mechanism for long-range correlations in DNA sequence is the most important issue both in physics and biology[1,2,3]. The purpose of this study is to understand universality of long-range correlations and the relation between long-range correlation and molecular evolution. There are two major theories for the origin of intron[4]. First, the exon theory of genes often called the intron-early view[5], second, the insertional theory of intron origins known as the intron-late view[6]. The intron-early view suggests that each of the exons is independent of the gene, each of the introns is a spacer, and that the present gene is assembled from their ancient genes. In contrast, with the intron-late view the current gene arises from intron-less gene by insertion of introns. Previous studies have shown that non-coding sequences such as intron display long-range correlation. But without regard to *the intron-structure*, the relation between the origin of intron and these long-range correlation has not been made clear in previous studies[1,2]. Therefore from the view-point of molecular evolution, we notice *the intron-structure*. If intron was an ancient spacer as thought in the intron-early view, *the intron-structure* of each of genes generally has no correlation. We showed that *the intron-structure in a single gene* has long-range correlation in all classifications, universally. Our results support the intron-late view. A existence of this universality leads to a thought that the intron-structure is advantageous for molecular evolution. We discuss such a molecular evolution to create universal long-range correlation.

2. Results and Discussion

Here, to select the sequence-date, we set up the following conditions. 1)Select a single gene data containing complete exon and intron sequences. 2)Select a single gene data which contains 3 or more introns, to quantify the arrangements of introns. Under these conditions we obtain 70 sequences from GenBank release 87, and the sequence-data covered 7 phyla and Plantae,

Fungi. To translate *the intron-structure* into the numerical equal base sequence $x_{m,\alpha}$, the method described below as considered,

$$
\begin{aligned}
x_{m,\alpha} &= 1 & &\text{if there is the base type } \alpha \text{ in the } m\text{th site} \\
& & &\text{and the } m\text{th site is in the region of intron,} \\
&= 0, & &\text{otherwise,}
\end{aligned} \tag{1}
$$

$m = 0, 1, ..., M - 1$, Where $x_{m,\alpha}$ is the numerical sequence of α type base in mth site. α is each of the base type A or C,G,T. M is the length of sequence(base-pairs,bp). To quantify the correlation of this numerical equal base sequence, we used Fourie transformation[3]. The equal base power spectrum $S_\alpha(f_n)$ can be expressed as

$$
S_\alpha(f_n) = |M^{-\frac{1}{2}} \sum_{m=0}^{M-1} x_{m,\alpha} \exp(-i2\pi f_n m)|^2,
$$

where f_n is frequency(bp^{-1}) defined as

$$
f_n = n/M, n = 0, 1, ..., \frac{M-1}{2}
$$

To reduce the influence that the difference of each of the base ratio, 4 type equal base spectrums in the same gene sequence are averaged as follows,

$$
S_{ave}(f_n) = \frac{1}{4} \sum_{\alpha=A,C,G,T} S_\alpha(f_n),
$$

where $S_{ave}(f_n)$ is the average spectrum.

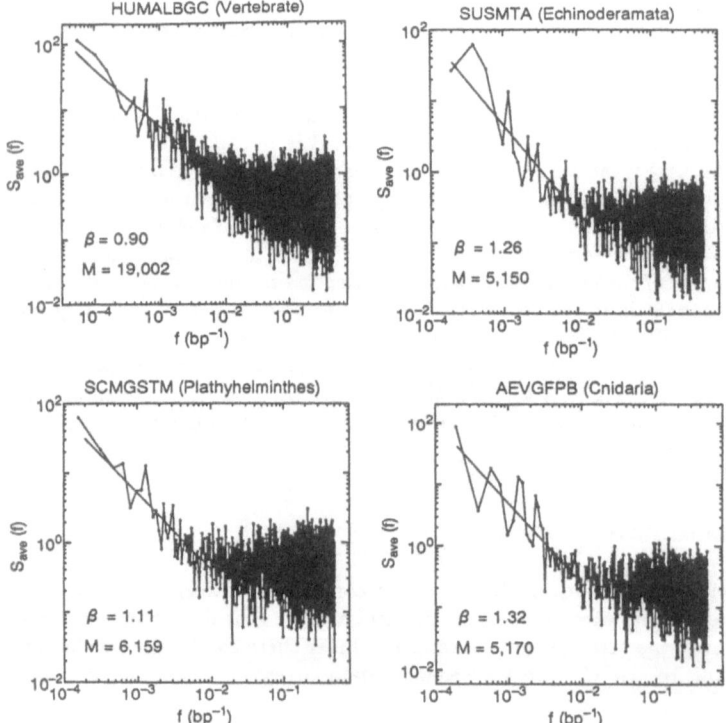

Fig.1 Double logarithmic plots of the average power spectrum $S_{ave}(f)$ of the single gene intron-structure in 4 classifications. From left to right, top to bottom, the figures are for HUMALBGC, SUSMTA, SCMGSTM, AEVGFPB. M is the DNA sequence length (base-pairs; bp). Least square fits by $1/f^\beta$ for $S_{ave}(f)$ components below $10^{-2}(bp^{-1})$ are shown as the solid lines. $S_{ave}(f)$ is remarkably similar in all. All of the β readings are almost 1.00.

Figure 1 shows typical average spectrums $S_{ave}(f)$ in each of the classifications. The increase of power at low frequency is observed in all classifications. The behaviour of every $S_{ave}(f)$ is, remarkably, almost the same. The solid lines are obtained from power law fitting for $S_{ave}(f)$ components below $10^{-2} bp^{-1}$. The exponents β of the fitting line $1/f^\beta$ are always 1.0. The classification-average exponents β_{ave} are almost constant at 1.0 in all classifications. We thus show that the behaviour of $S_{ave}(f)$ is all similar. We find that β_{ave} is about 1.0 in all classifications. In contrast with previous studies, we don't find any evolutionary change in β_{ave}[3]. To confirm the range of these long-range correlations, we investigate sequences containing more than two genes. Figure 2 shows $S_{ave}(f)$ of HUMHBB(Vertebrata) containing 6 genes. Low frequency components corresponding to correlation extended over more than 2 genes do not show power-law behaviour. These spectrums do not show long-range correlations over more than 2 genes.

These indicate that *the intron-structure* in every single gene has no random arrangement or gene-dependent structure, but scale invariant character or fractal property. Since the ancient spacer-structure depend on species and gene-type, our results are not consistent with the intron-early view. Therefore, our results support the intron-late view. It can be thought that the intron insertion system in *a single gene* is the origin of long-range correlations. From the intron-late view, our results indicate that the intron-structure evolved not in an accumulation of accidentally mutations, but to a direction of self-organization. This idea will show a new aspect of neutral theory in evolution[7]. Universal feature of our results implies that evolution by random mutation transforms directional at a certain time. It implies that such a system is related to self organized criticality(SOC). The idea of SOC is the common explanation of the origin of fractal shape or 1/f spectrum in complex systems, such as a biological system[8]. To understand this system has more important influences on biology and physics.

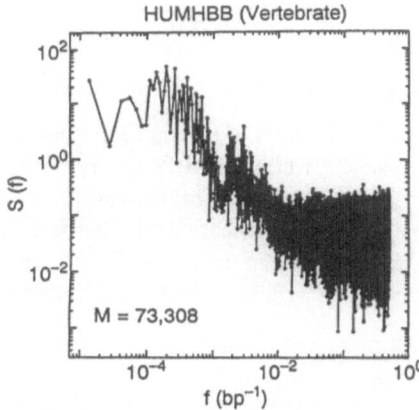

HUMHBB (Vertebrate)

Fig. 2

Double logarithmic plot of the average power spectrum $S_{ave}(f)$ of non-single gene intron-structures. The figures are for HUMHBB(Vertebrata) containing 6 genes. Low freqency components corresponding to correlation extended over more than 2 genes do not show power-law behaviour. This indicates that long-range correlation of the intron-structure is only effective in a single gene.

References

[1] C.-K.Peng, S.V.Buldyrev, A.L.Goldberger, S.Havlin, F.Sciortino, M.Simons. and H.E.Stanley, Nature **356**, 1992, pp.168-170.

[2] W.Li and K.Kaneko, Europhys. Lett. **17**, 1992, pp.655-660.

[3] R.F.Voss, Phys. Rev. Lett. **68**, 1992, pp.3805-3808.

[4] L.D.Hurst, Nature **371**, 1994, pp.381-382.

[5] W.F. Doolittle, Nature **272**, 1978, pp.581-582.

[6] W.Gilbert, Nature **271**, 1978, pp.501-505.

[7] M.Kimura, *The neutral theory of molecular evolution* Cambridge Univ.Press, 1983.

[8] P.Bak and K.Sneppen, Phys. Rev. Lett. **71**, 1993, pp.4083-4086.

Simulation of the oscillatory potential propagation in the slug's brain with an array of nonlinear oscillators

Atsushi Yamada, Tetsuya Kimura, Eiji Kono and Tatsuhiko Sekiguchi

SANYO Electric Co., Ltd., Tsukuba Research Center, Tsukuba-shi, Ibaraki, 305 Japan

Abstract

We simulate an oscillatory propagating waves in the olfactory center (procerebrum:PC) of a terrestrial slug, *Limax marginatus*, with an array of nonlinear oscillators. We search the conditions where the entrained waves are constantly propagating from one terminal of the array to the other one. As regards the interaction of the oscillators, the inputs which are proportional to the differences between amplitudes of oscillators next to each other are needed for constant propagation. In addition to the condition, the oscillators are necessary to be arranged regularly according to their proper frequencies. We can simulate the PC with an array of nonlinear oscillators in the way which is consistent with the morphological and physiological measurements such as gap junctions or frequency gradient.

Key words: slug olfactory system, Van der Pol oscillator, gap junction

1. Introduction

The central nervous system of the slug, *Limax marginatus*, consists of several ganglia including a pair of cerebral ganglia. PC is a part of the cerebral ganglion which receives olfactory information from the tentacle nerve. We can observe the propagating waves of the local field potential(LFP) in the PC with electrical [1] and optical [2,3] methods. The direction of propagation is always from the lateral end to the medial end. Even if the PC is cut into small pieces, these pieces can continue to oscillate electrically. This result suggests that the PC consists of an array of nonlinear oscillators.

2. Experimental

Searching the mechanism of the propagation, we simulated the propagation of oscillatory potential in the PC with a one-dimensional array of nonlinear oscillators such as Van der Pol which is characterized by the following difference equations:

$$\ddot{x} + a_n(x^2 - 1)\dot{x} + x = d_n \tag{1}$$

where a_n is the nonlinear parameter which decides on the proper frequency of the nth oscillator, d_n is the external force of the nth oscillator. Interactions between the PC cells were known so little that we assumed four types of plausible interaction. We tried to simulate with four types of interactions between oscillators correspondingly. First was that one oscillator received inputs which were proportional to the differences between the amplitudes of oscillators next to each other. This interaction is given by

$$d_n = k((x_{n+1} - x_n) + (x_{n-1} - x_n))$$

(2)

and assumed that there were gap junctions between PC cells physiologically. Second was similar to the first one, but one oscillator received inputs which were proportional not to the differences but to the amplitudes itself of the next oscillators. This interaction is given by

$$d_n = k(x_{n+1} + x_{n-1})$$

(3)

and assumed synaptic connections between PC cells physiologically. The third was that every oscillator equally received the input which was proportional to the summation of every amplitude. This interaction is given by

$$d = k \sum_n x_n$$

(4)

and assumed the integrated cell which received inputs from every cell. The fourth was that one oscillator received inputs which were exponentially inverse-proportional to the distance between oscillators, and were directly proportional to the amplitudes of them. This is given by

$$d_n = k \sum_{m(m \neq n)} 2^{-(m-n)} x_m$$

(5)

and assumed diffusive interaction such as hormonal one.

Under these conditions, we arranged the array of twenty oscillators in several ways. We simulated it without interaction by time 100. Since then, we simulated it with interaction and checked whether entrained waves could propagate or not. We represented the results in the manner where a bar (|) meant the maximum amplitude and a dot (.) meant the minimum amplitude shown as Fig. 1.

3. Results and Discussions

As regards the interaction, entrained waves could propagate only in the manner where the inputs were proportional to the differences between the amplitudes of oscillators next to each other as given in (2). As regards the arrangement, entrained waves could propagate constantly through the array nothing but in the manners below. One was the manner which the oscillators are regularly arranged according to their proper frequencies. The other was the manner which the oscillator with the maximum proper frequency is at one end of the array, and other oscillators were identical. In these manners, propagating waves were constantly formed from the terminal oscillator with higher proper frequency to the other terminal (Fig. 1). Entrained frequency was almost equal to the maximum proper frequency. These results suggest that the PC consists of oscillators which are connected next to each other on the above conditions. Several physiological data are consistent with this simulation.

When we applied the PC to a type of stimuli such as olfactory, gustatory, electrical or pharmacological one, we could observe that both the frequency of the oscillatory LFP and the velocity of the LFP propagation were changed. The modulation of the frequency was, however, not related to that of the velocity [4]. This result is consistent with the simulation in which the entrained frequency depends on the oscillator with the maximum proper frequency, while the velocity depends on the strength of the connection between the oscillators [5].

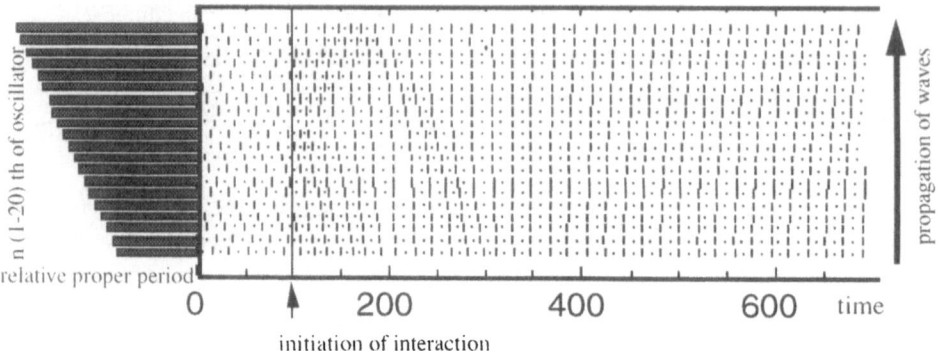

Fig. 1. Simulated result in the manner where the oscillators are regularly arranged according to their proper frequencies. After the initiation of interaction, every oscillator is gradually getting entrained. At last, the propagation of entrained waves is formed from the terminal oscillator with higher proper frequency (lower) to the other terminal (upper).

4. Conclusion

To form the propagation of entrained waves constantly through the array of oscillators, it is necessary that the interaction was proportional to the differences between the amplitudes of oscillators next to each other, and that the manners of arrangement were according to proper frequency regularly or there is a special oscillator with higher proper frequency at one terminal. We can simulate the PC with an array of nonlinear oscillators in the way which is consistent with the physiological measurements.

5. Acknowledgment

The present study was supported by Special Coordination Funds of the Science and Technology Agency of the Japanese Government.

6. References

[1] T. Kimura, A. Yamada, H. Suzuki, E. Kono, T. Sekiguchi, Y. Sugiyama, *Olfaction and Taste IV*, 1994, pp. 440.
[2] D. Kleinfeld, K.R. Delaney, M.S. Fee, J.A. Flores, D.W. Tank, A. Gelperin, *Journal of Neurophysiology*, Vol. 72, 1994, pp. 1402-1419.
[3] T. Kimura, S. Toda, S. Kawahara, T. Sekiguchi, Y. Kirino, *in preparation*
[4] A. Yamada, T. Kimura, T. Sekiguchi, *in preparation*
[5] Y. Kuramoto, *Progress in Theoretical Physics*, Vol. 71, 1984, pp. 1182-1196.

Autonomous Control of Heartbeat as Entrainment between Mechanical Environment and Electrical Rhythms

Yosuke Ariizumi, Yasuaki Kurita and Yoko Yamaguchi

Dept. of Information Sciences, Tokyo Denki Univ., Hatoyama-machi, Saitama, JAPAN

Abstract

We elucidate how the heartbeat is controlled under various changes in mechanical environments. The ventricular model used here includes processes of the electrical rhythm, muscle contraction and the blood flow. They are coupled by mechanoelectrical feedback (MEF) as well as by excitation-contraction coupling. Our results show the diversity in dynamics of the heartbeat, various limit cycles and chaos. Under the condition of the generation of the limit cycle, the heartbeat is stable during changes in the venous pressure. We conclude that heartbeat is autonomously stabilized by self-organization of temporal coherence through MEF.

Keywords: arrhythmia, ventricular model, chaos, mechanoelectrical feedback, self-organization

1. Introduction

The heartbeat is always exposed to changes in mechanical environments such as the change of velocity or pressure of blood flow. Since they continuously change either periodically or nonperiodically, the temporal relation of electrical and mechanical processes significantly contribute to the efficiency of the heartbeat. The coupling between the two processes has usually been considered as the one-way action of excitation-contraction coupling. The contribution of the opposite way from the mechanical to the electrical process is rarely considered in heartbeats. Recently it was shown that the stretch activated ionic currents (i_{SAC}) significantly contribute to the mechanoelectrical feedback (MEF) (Lab & Holden, 1991). The present paper is devoted to elucidation of autonomous control of the heartbeat by self-organization of temporal correlation between the two processes under various changes of mechanical environments.

2. A Theoretical Model of the Heartbeat with MEF

We use the model of the heartbeat including electrical and mechanical processes proposed by Hori et al. (1994). The model consists of electrical, muscle contraction and blood flow processes in a ventricle cylinder as shown in Fig. 1. The electrical process is given by the membrane potential, ionic channels and calcium concentration according to the Beeler Reuter equation (BR eq.). The muscle contraction process is described by the calcium-tension, length-tension and tension-velocity

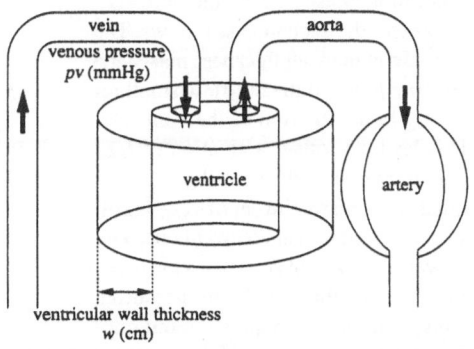

Fig. 1 Schematic structure of the heatbeat model

relations. The blood flow is described by the pressure and volume of the cylinder and in-flow from the vein and out-flow to the aorta. i_{SAC} is introduced to the BR eq. as a volume-dependent current that is a mediator of MEF. For the sake of simplicity, the electrical and mechanical quantities are assumed to homogeneous in a cylinder. The above model is given by 11 dimensional differential equations. Numerical analyses of the model were carried out by the Runge-Kutta-Gill method.

3. Heartbeat under Various Values of Thickness of Ventricular Walls

We first analyzed the contribution of the mechanical environment to the heartbeat under various values of thickness of ventricular walls, w, by computer experiments using the above model. Figure 2 shows typical examples of the time evolution of the membrane potential (E) and i_{SAC}. It is found that regular rhythms of E and i_{SAC} are generated in a normal condition, as shown in Fig.2a, while multiperiodic or nonperiodic activities appear in conditions of values of thickness less than the normal one, as shown in Figs.2b,c. Further decrease in the thickness brings about the stable equilibrium point. Thus, thinning of walls leads the heartbeat to what is called the bigeminy, arrythmia and cardiac arrest. These drastic changes of the heartbeat under various values of the thickness are obviously brought on the MEF. On the other hand, the MEF itself depends on the electrical activities. The mutual interactions between the two processes give rise to the variety in their relations. In a normal condition (Fig.2a), E and i_{SAC} are mutually entrained as a harmonic synchronization. In the case of bigeminy (Fig. 2b), they are entrained with the winding number, 2: 1. In the case of non-periodic motion (Fig. 2c), the Lyapunov exponent is found to be positive. That is, a chaos attractor is generated by these interactions resulting in the arrythmia. It is discovered that in the above bifurcation, the physiological role of i_{SAC} qualitatively changes. In the case of a normal condition, i_{SAC} is almost inward and synchronized in phase with the inward natrium current. This means that i_{SAC} is cooperative with the excitation of the membrane potential. On the other hand, in the case of the thin wall, i_{SAC} suppresses the excitation as follows. The decrease in the wall thickness makes the pressure low and the out-flow small resulting in the dilation of the ventricle. This leads to longer activation of i_{SAC} than in the case of the normal one. In this situation, inward current of i_{SAC} is followed by a significant outward one. The outward component suppresses the depolarization of the membrane potential. Thus, i_{SAC} is not cooperative but conflicts with the excitation of the electrical rhythm.

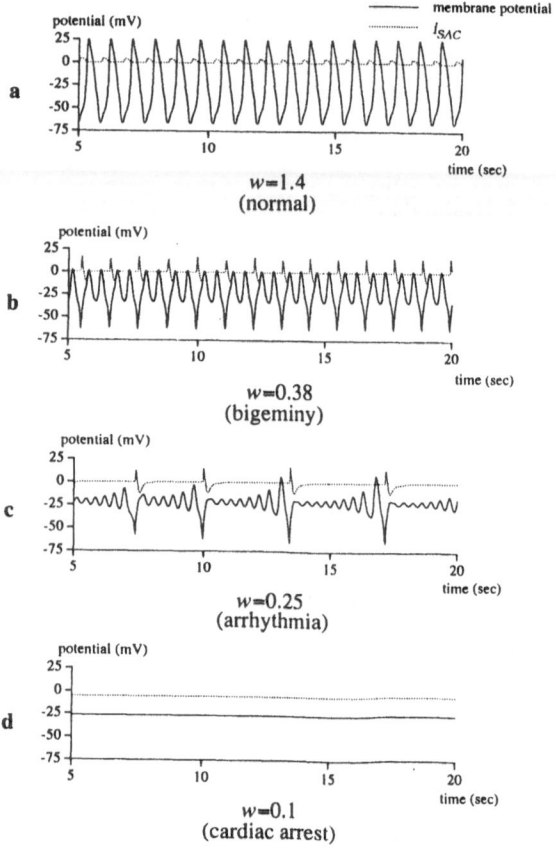

Fig. 2 Membrane potential and i_{SAC} for various values of w

4. Heartbeat under Varying Venous Pressure

For the purpose of examining the effect of the change of blood pressure, the venous pressure, p_v, was given by various fixed values or by the sinusoidal changes. In computer experiments, both the case with i_{SAC} and without i_{SAC} were elucidated to facilitate comparison. It should be noted that either the stroke volume or the period remains constant in the range of $p_v > 50$ (mmHg) for various fixed values of p_v. In the absence of MEF, it is seen that the continuous change of p_v directly brings the continuous and irregular change of the period for the sinusoidal changes of p_v. On the contrary, in the presence of MEF, the heartbeat is regularly rhythmic by the period of p_v. This means that the heartbeat is entrained by the continuous change of p_v to lead a kind of stable stationary state. This stability is obtained in the range of the amplitude of $p_v <95$(mmHg) .

5. Summary and Conclusion

We have shown that the heartbeat is regularly rhythmic due to the entrainment between the electrical and mechanical processes through MEF in normal conditions. This entrainment is stable against various changes of physiological parameters to enable autonomous control near normal conditions. However, change beyond a critical value finally disrupts the entrainment. Thus, arrythmia appears as chaos due to the failure of entrainment. Though our model is a simple one, these results suggest important issues: the entrainment through the MEF essentially contributes to autonomous control near normal conditions, and the disruption of the entrainment by the MEF brings on the significant damage to the heartbeat.

Acknowledgment: The authors express their appreciation to Mr. S. Hori and Prof. H. Shimizu for their valuable discussion and cooperation and to Prof. Davis for improvement of English.

6. References

[1]   G. W. Beeler, H. Reuter, Reconstruction of the action potential of ventricular myocardial fibers, *J. Physiol*, **268**, 1977, pp. 177-210.

[2]   M. Akay, W. Craelius, Mechanoelectrical feedback in cardiac myocytes from stretch-activated ion channels, *IEEE Transaction on biomedical engineering*, **40**, 1993, 811-816.

[3]   M. J. Lab, A. V. Holden, Mechanically induced changes in electrophysiology: Implications for arrhythmia and theory, *Theory of Heart*, Springer-Verlag, 1991.

[4]   M. R. Franz, R. Cima, D. Wang, D. Profit, R. Kurz, Electrophysiological effects of myocardial stretch and mechanical determinants of stretch-activated arrhythmias, *Circulation*, **86**, 1992, 968-978.

[5]   S. Hori, Y. Yamaguchi, H. Shimizu, Self-organization of heartbeat through mutual entrainment among the myocardial cells and blood flow, *Proceedings of Bioengineering and physiological engineering symposium*, 1994, pp. 355-358. (in Japanese)

[6]   Y. Ariizumi, S. Hori, Y. Yamaguchi, H. Shimizu, Arrhythmia as a chaos induced by Mechanoelectrical feedback, Proceedings of Life Support Society, 1995, pp.63 (in Japanese)

Chaotic behavior of the Hodgikin-Huxley equations under small random noise

Hiroaki Tanaka and Kazuyuki Aihara

Dept. of Mathematical Eng. and Info. Phys., Faculty of Eng., The Univ. of Tokyo, 7-3-1 Hongo, Bunkyo-ku, Tokyo 113, JAPAN

Abstract

The chaotic behavior of the neuron is studied by calculating Hodgikin-Huxley equations under white noise with various amplitude. Even small noise, such as 10^{-10}, effects the chaotic behavior. With noise from 10^{-10} to 10^{-1}, the trajectory of the strange attractor converges in a bi-phasic manner. With large noise, the trajectory shows some macroscopic periodicity, which may be accompanied with its shortening of the period. The amplitude dependence of the noise effect on the chaotic behavior can be explained by two types of scaling structure of the nerve chaos.

Key words: Hodgikin-Huxley equations, noise, order

1.Introduction

The excitable membrane potential of the squid giant axon shows chaotic behavior under certain sinusoidal current stimulation[1,2,3]. Matsumoto and Tsuda[4] showed that a certain amount of noise, such as 10^{-4}, induced some order in the B-Z chaos. In the actual neuron or the physiological experiment, it is exposed to the certain amount of thermal and/or electrical fluctuation (i.e. noise). Thus, the observed chaos in the neuron may be affected by such noise. The chaos in the neuron can be well simulated by the numerical calculation of Hodkin-Huxley equations[5]. In this paper, the effect of the noise with various amplitudes is studied numerically with the Hodgkin-Huxley equations.

2.Simulation

Detail methods and parametric conditions for the numerical calculation are based on Usami et.al.[5]. White noise with various amplitudes is superimposed to the sinusoidal current stimulation term. To ensure the accuracy of the calculation, the noise amplitude more than 10^{-10} is used for the analysis.

2.1 Wave form

By comparing the chaotic wave forms of the membrane potential with various noise amplitudes, it is found that even at the small noise as 10^{-10} the wave form is occasionally changed from that without noise. That is, the small noise as 10^{-10} may affect the nerve chaos, although the appearances of the wave forms are not different from that without noise. Obvious difference can be observed in the wave forms with larger noise than 10^0. The similar sequence of peaks with almost same periodicity is observed frequently, suggesting some macroscopic order. This macroscopic periodicity seems to be accompanied with the shortening of the period.

2.2 Peak height analysis

To characterize the chaotic behavior against the noise amplitude, distribution of the peak height of the wave forms of the membrane potential is analyzed. Without noise, five major peaks and 11 minor peaks are observed in the peak height distribution.

At the noise amplitude of 10^{-10}, although the wave form can not be distinguished from that of no noise, the 11 minor peaks in the distribution increase more than twice than those of no noise, while the major peaks are not affected. The increase of the 11 peaks can be observed at the noise amplitude up to 10^{-3}. This increase of the peaks is caused by some convergence of the trajectories. We call this phenomena of the increase of the peaks as phase 1.

Above 10^{-2} to 10^{-1}, 11 minor peaks decrease and disappear, while some of the major peaks increase. This seems that the converged trajectories at phase 1 are diverged, while another type of convergence occurred (phase 2).

Above 10^{-0}, all the peaks become lower with broadening the peak height distribution. This seems whole divergence in the trajectory is caused by large noise (phase 3).

2.3 Return map and peak sequence

From the peaks of the wave forms, return-maps are obtained for the various noise amplitudes. It was found that fundamental structure of the return map is not changes by the noise, although the fluctuation of the points becomes conspicuous especially at larger noise amplitudes (phase 3).

The five major peaks and the 11 minor peaks are related as the mapping each other, respectively. In addition, the increase of the 11 minor peaks can be explained by the increase of the points located at the specific region on the diagonal of the map. The points of the region at phase 1 increase about 12% than without noise, while they decrease significantly with increasing the noise amplitude (phase 2, 3). Analysis on peak sequence on the map also suggests that some change of fine structure in the mapping causes phase 1.

At phase 3, the points relevant to the macroscopic periodicity increase, while some other points decrease. The peak sequence analysis also shows the increase of short term wandering on the map.

2.4 Noise induced trajectory separation

To examine the effect of the noise in detail, the trajectories with some noise and without noise are calculated simultaneously with checking the distance between those in the phase space. It is found that both trajectories usually evolve in a same manner, but occasionally separate each other. Thus, the number of the separation for the same time duration is counted for the various noise amplitudes.

It is found that log-log plot of the number versus the noise amplitude shows bi-phasic linear relation. The scaling factors are estimated on the basis of the following relationship:

$$N \propto r^D$$

where, N is the number of separation points, r is the noise amplitude and D is the scaling factor. Thus, the scaling factor measured by the noise induced trajectory separation are 0.075 for phase 1 and 0.3 for phase 2 and 3. This can be explained by the existence of two types of the scaling structure.

The plot of the point of the separation in the phase space shows that the points are localized along certain trajectory for phase 1, while they seem to blur at phase 2 and 3.

3.Discussion

In this study, we find that even small noise, such as, 10^{-10}, may effect the nerve chaos essentially. The convergence of the trajectory is observed. This noise amplitude is extremely smaller than the amplitude of noise-induced order (NIO) reported by Matsumoto and Tsuda[4].

Matsumoto and Tsuda[4] discussed that the reason of NIO is the balance of the trajectory convergence and divergence induced by the trajectory transition by noise. The results of this study can be explained by their mechanism basically, although two scaling structure is involved.

The effects of the noise are summarized as follows:

Phase	Noise amplitude	Scaling structure 1 with factor: 0.075	Scaling structure 2 with factor: 0.3	phenomenon
1	$\leqq 10^{-3}$	convergence	no effect	increase of the minor peaks in the peak height
2	$10^{-2} \sim 10^{-1}$	divergence	convergence	increase of some major peaks in the peak height
3	$\geqq 10^{0}$	divergence	divergence	macroscopic order

References

[1] K. Aihara, G. Matsumoto, Y. Ikegaya, Periodic and non-periodic responses of a periodically forced Hodgkin-Huxley oscillator, *J. theor. Biol.* Vol. 109, 1984, pp. 249-269.

[2] K. Aihara, G. Matsumoto, M. Ichikawa, An alternating periodic-chaotic sequence observed in neural oscillators, *Phys. Lett. A*, Vol. 111, 1985, pp. 251-255.

[3] K. Aihara, T. Numajiri, G. Matsumoto, M. Kotani, Structures of attractors in periodically forced neural oscillators, *Phys. Lett. A*, Vol. 116, 1986, pp. 313-317.

[4] K. Matsumoto and I. Tsuda, Noise-induced order, *J. Stat. Phys.* Vol. 31 , 1983, pp. 87-106.

[5] T. Usami, T. Yamada, N. Ichinose, K. Aihara, Hodgikin-Huxley equation and its response to the periodic stimulation, *J.SICE*, Vol.34, 1995, pp. 769-774.

A Reduction Model for Breaking Phase-Order Preservation in Globally Coupled Oscillators

Takeshi Takaishi[1] and Yasumasa Nishiura[2]

1 Faculty of Engineering, Hiroshima-DENKI Institute of Technology, Hiroshima , 739-03
2 Research Institute for Electronic Science, Hokkaido University, Sapporo, 060

Abstract

A new type of global interaction called "active" type is introduced to the pulse-coupled oscillators. The order preserving property in phase space breaks down under this interaction, i.e., it frequently happens that one oscillator passes another in phase space. This makes the dynamics richer than the passive case, in fact, many periodic orbits appear besides the synchronous state.

Keywords: Coupled oscillator, pulse-coupled system, phase dynamics

1 Introduction

Mirollo and Strogatz proposed a simple reduction model of phase dynamics with the reset-effect of the pulse-coupled oscillator system with all-to-all interaction [1]. They reduced its interaction of the system, such that its instantaneous interaction is regarded as the increment of the certain amounts of its internal voltage. Nishiura et al. find the hierarchy of clustered-periodic solution in the system with inhibitory coupling [2]. But they only adopted its results on the systems with the homogeneous interaction, where the strength and the sign are equivalent in all interaction.

Through the results found by Mirollo and Strogatz, or Nishiura et al. it shows interesting features, but the diversity in the original nonlinear system does not appear in those. The main reason is that these systems keep their order of phase at the initial state for $t > 0$. We call this order-preservation system. We make a simple approach to appearance of the diversity with changing the properties of their interaction, where the initial order of phase can not hold any more and the variety periodic solutions appear. It generically exists in the nonlinear system with mutual interaction.

The main aim of this report is to introduce a new type of interaction among oscillators although we employ a phase dynamics model like Mirollo and Strogatz's, that is defined its interaction on the effected / effectant side, non-firing/firing one, the *active*-type and the *passive*-type. We find that the model of *active*-type (*active*-model) shows the diversity of periodic solution though the model of *passive*-type (*passive*-model) have the restriction about periodic solution.

2   Model and Simulation Results

For simplicity we treat the Mirollo and Strogatz's solvable model below [1] . When the
voltage excesses the threshold, 1, as the result of the other's firing, the voltage is recognized
to reach threshold. Then one release the accumulated voltage and immediately the phase
becomes 1 to 0. This oscillator synchronize to the firing one with the *reset* of phase. If some
synchronized cluster of oscillators fire, the other oscillators change their voltage as the effect
piled up.

Here we introduce the "*Active*-type" in the interaction of pulse-coupled system, that the
strength of the interaction, or voltage variance, is defined by the side of the effected oscillators, non-firing one. This definition allows oscillators to pass the other in phase ring. And
we also call the another system, as "*Passive*-type" system in the contrary with the system
of *active*-type.

Now, we show the difference between these 2 types of interaction on 3-oscillator system,
(+ − −), where the strength of interaction are characterized as followings: No. 1 is positive
interaction and No.2 and 3 are negative interactions.

Figure 1 shows the concepts of *active*- and *passive*-model of the 3-oscillator system. Let us
consider the oscillator No.1 (positive), No.2 (negative), and No.3 (negative) in ther reverse
order of phase at the initial. As time goes, the oscillator No.3 reaches the threshold level and
fires. In the system with the *passive*-type, the oscillators No.1 and No.2 change their voltage
smaller, which is defined by the oscillator No.3. In the system with the *active*-type, the
oscillator No.1 changes its voltage larger and the No.2 changes it smaller when the oscillator
No.3 fires, the strength of the interaction are defined by the oscillator No.1 and 2. When the
No.3 fires the oscillator No.1 passes the No.2 in phase if the initial phase difference between
oscillator No.1 and 2 is sufficiently small. It could not appear in the *passive*-system.

In the *active*-system we can observe many periodic solution according to the initial phase
distribution. *Figure* 2 shows the computational result that is the classification of initial
phase distribution of ϕ_2 and ϕ_3 ($\phi_1 = 0$ at the initial). In the passive-system with the same
system parameter, we can only find the sychronous solution (period = 0).

Figure 3 is the transition diagram of (+ − −) system. In the *passive*-system, 3 oscillators
fire according to the initial order, however, the *active*-system has some by-pass routes and
((+−)(−)) collapsed state (hashed lines) more. The collapsed state is that a positive oscillator and a negative oscillator synchronize for some interval till another one fires. From the
computational results, we find that the selection of route between states is firmly related to
the initial state, and that the by-pass routes and the collapsed state emerge the diversity of
periodic solution.

3   Summary

We compare 2 types of the reduction model of pulse-coupled oscillator system. which is
improved from the Mirollo and Strogatz's. We show that the *active*-system shows the various

periodic solution because of the lacking of its order-preservation property. In this system, the diversity of the periodic solution which the coupled system originally has are recovered by the rule of the interaction, not by the asymmetry of the interaction.

References

[1] R.E.Mirollo and S.H.Strogatz, Synchronization of pulse-coupled biological oscillators, *SIAM J.Appl.Math.* **50** (1990), 1645-1662.

[2] Y.Nishiura, J.Shidawara, and T.Takaishi, Dynamics of inhibitory pulse-coupled oscillators, *Dynamical Systems and Applications 4* (1995), 549-561, World Scientific,

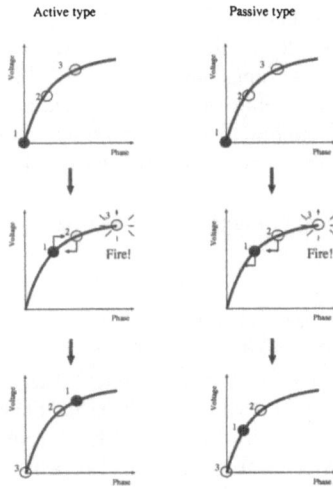

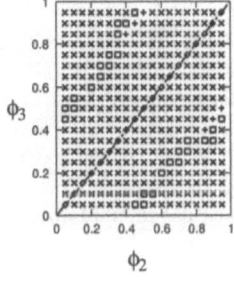

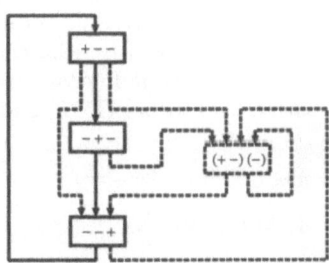

Figure 1: Firing of the *Active*- and *Passive*-type for $(+ - -)$ 3 oscillators system.

Figure 2: The distribution of periodic solutions for initial phase (ϕ_2, ϕ_3) with $\varepsilon = 0.1, b = 1$. (In the initial state we assume $\phi_1 = 0$.) ◇, +, □, × describe the periodic solutions with period = 0 (synchronous state), 8, 15, 22.

Figure 3: Transition diagram for periodic solutions in $(+ - -)$ system. We start from the $(+ - -)$ state. Hashed lines describe the route and the state that are existed only in *Active*-type.

Critical phenomena in lattice Lotka-Volterra model: Parity law and selfstructuring extinction pattern

Kei-ichi Tainaka, Kiyoshi Kobayashi

Dept. of Physics, Ibaraki Univ., Mito, 310, JAPAN

Abstract

So far, we have developed a lattice version of Lotka-Volterra theory (LVT), and found that long-term forecast looks unpredictable. However, in this article, we find a parity law: Not a microscopic structure of ecosystem, such as detailed interactions between species, but "parity", defined by whether each system has odd or even number of species, is relevant for the long-term forecast. This parity dependence can be explained by the combination of mean-field theory (LVT) and the effect of spatial pattern formation of endangered species.

Key words: Lattice ecosystem, critical phenomena, endangered species, parity law

1. Introduction

By the use of computer, we mainly study "self-response": we permanently apply a perturbation (such as the change of value of birth or survival rate) to a target species, and record the long-term response of the target species. For example, a birth rate of a target species is increased by a perturbation. The system is assumed to stay in a stationary state before the perturbation. Just after the perturbation. the abundance of the species always increases (short-term response). However, later, it sometimes decreases in a new stationary state. In the below, the term "opposite response" is used, when the long-term response is just opposite to the short-term one. The purposes of the present paper is to present a parity law which is useful for the long-term forecast in lattice models. and it enables us to know when the opposite (counterintuitive) response occurs.

2. A Model

We first consider the following cyclic model:

$$X_{i-1} + X_i \xrightarrow{r_i} 2X_i, \tag{1a}$$

$$X_n \xrightarrow{d} X_0, \tag{1b}$$

where X_i means an individual of consumer i ($i = 1, \cdots n$), and X_0 represents the producer. The parameter r_i is the reproduction rate of X_i, and d denotes the death rate of species n. We put $r_1 = r$ and $r_i = 1$ ($2 \leq i \leq n-1$). Since X_i eats X_{i-1}. the species n denotes the top predator; nevertheless. when X_n dies, it is served for the producer X_0. Hence, the system (1) represents a food cycle composed of $n + 1$ kinds of species ($2 \leq n \leq 7$). When n is even (odd), "parity" of this system becomes odd (even). where the parity is defined by whether a system has odd or even number of species.

In this article, we apply a lattice version of LVT [1-3]: the two-body reactions (1a) occur between eight neighboring lattice sites. If $n = 1$ and 2. our system is called the contact process [4] and prey-predator system [1, 5], respectively. The population dynamics of lattice model is stabilized into a unique state depending on the values of parameters. Irrespective of initial patterns, the system evolves into a stationary state, where the configuration of spatial pattern varies with time.

In Fig. 1, the steady-state density of species 1 is plotted against its birth rate r, where we fix the parameter d. This figure shows the following results: i) If n is odd (parity is even). the long-term response has the same direction as the short-term one: the steady-state density of species 1 increases with increasing r. ii) in the case of odd parity, the opposite response occurs; the species 1 conversely decreases with increasing r. In Fig. 2, we depict the density of species 0 against the parameter d which denotes the birth rate of species 0. We also find that when the parity of the system is even (odd), both short- and long-term responses point to the same direction (opposite directions). The mean-field theory (MFT = LVT) [6] indeed explains the parity dependence in Fig. 2, but never predict the opposite response in Fig. 1. This discrepancy between lattice model and MFT becomes clear in the region where r takes small values: note that this region denotes the critical phenomena near extinction.

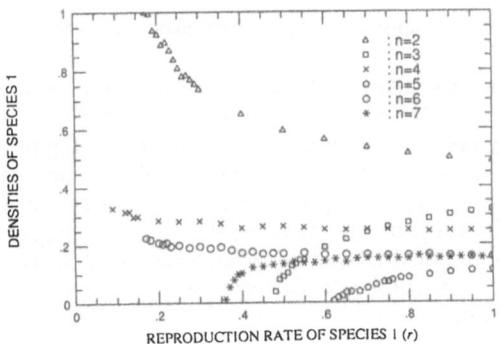

Fig. 1 The steady-state density of species 1 is plotted against its reproduction rate $r_1 = r$ ($d = 0.6$). Each plot is obtained by the long-time average in the stationary state ($200 < t \le 1000$) with 100×100 lattice.

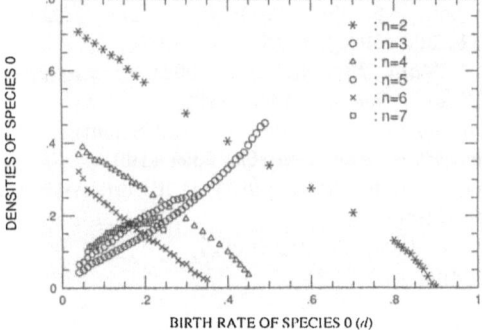

Fig. 2 Same as Fig. 1, but the density of species 0 is depicted against its birth rate d ($r = 1$).

3. Selfstructuring Extinction Pattern

As shown in Fig. 1, MFT fails for the critical phenomena near extinction. This failure comes from "selfstructuring extinction pattern (SEP)"of endangered species [1-3]; that is, when a *biospecies* faces extinction, the degree of contagiousness of the biospecies rapidly increases, where we call the biospecies, if it can be increased only by an autocatalytic reaction (reproduction process) [see the reaction (1a)]. The combination of MFT and SEP well explains the opposite response. For example, we consider the case of $n = 4$. When r decreases, the species 2 and 4 face extinction as predicted by MFT. Then, they are rapidly clumped. When r decreases and approaches extinction, the species 2 cannot easily eat its prey (species 1). Hence, the population size of prey increases in spite of the decrease of r.

4. Parity Law

Next, we check whether the parity dependence in system (1) holds for other lattice models. We analyze "cyclic systems"composed of only one loop of food chain. We find that if the system is situated far from extinction, the perturbation experiment (parameter dependence) can be qualitatively explained by MFT, whereas MFT fails for the critical phenomena. Our attention is, therefore, paid only to these phenomena. We find the following "parity law":

  i) When the parity of the system is even, no opposite response occurs except for one case [7] (among 20 critical phenomena).
 ii) When the parity is odd, the long-term response is very frequently opposite to the short-term one (71% for 24 cases).

Here, we omitted the ambiguous case that we cannot determine whether the population size of a target species increases or decreases by perturbation. Although MFT explains the result i), this theory never sufficiently predicts the result ii): the frequency of opposite response decreases (46%). We describe the above result ii) in detail. Our perturbation is classified into two groups; namely, the change of birth or death rate. In the former case, almost all responses are opposite. On the other hand, in the latter case, the opposite response never frequently occurs; this is due to the fact that when we strongly increases the death rate of a species ("overkilling"), the species always decreases and goes extinct. There is no opposite response for the overkilling.

References

[1] K. Tainaka and S. Fukazawa, *J. Phys. Soc. Japan*, **61**, 1891 (1992); K. Tainaka, *J. Theor. Biol.* **164**, 91 (1994).

[2] K. Tainaka and Y. Itoh, *Europhys. Lett.* **15**, 399 (1991); *Phys. Lett.* **A220**, 58 (1996); K. Tainaka, *Phys. Lett.* **A 176**, 303 (1993); **A 207**, 53 (1995).

[3] K. Sato, H. Matsuda and A. Sasaki, *J. Math. Biol.* **32**, 251 (1994); H. Matsuda, N. Ogita, A. Sasaki and K. Sato, *Prog. theor. Phys.* **88**, 1035 (1992).

[4] N. Konno, *Phase Transition of Interacting Particle Systems* (World Scientific, Singapore, 1994); R. Dickman and I. Jensen, *Phys. Rev. Lett.* **67**, 2391 (1991).

[5] B. Drossel and F. Schwabl, *Phys. Rev. Lett.* **69**, 1629 (1992); B. R. Sutherland and A. E. Jacobs, *Complex Systems* **8**, 385 (1994).

[6] For example, T. Chawanya, *Prog. Theor. Phys.* **94**, 163 (1995).

[7] R. Dickman, *Phys. Rev.* **B 42**, 6985 (1990).

Part V Social Systems

Complexity and Economics of Institutions

Toshiji Kawagoe

Faculty of Economics, Saitama University, Urawa-shi, Saitama 338, JAPAN

Abstract

In this article, we focus on computational aspects of game strategies and mechanism design. Some games or mechanisms arise the problems with computational difficulties. We argue some complexity measure of game strategy to model agents whose rationality are bounded. We think that the model of bounded rational agent is more suitable for analyzing many economic situations than that of rational agents assumed in ordinary economics. In fact, we can show that bounded rational agents can achieve Nash equilibrium in optimal auction mechanism. But human behaviors in real life have some mysterious aspects. Experiments with human-subjects showed that players could have some different expectations or actions in the game that could not be obserbed in Nash equilbrium analysis. Constructing more reasonable behavioral model of bounded rationality remains in future researches.

Key words: Bounded rationality, complexity measure, mechanism design, auction experiment

1. Introduction

Complexity considerations in economics come to be accepted by researchers in economic dynamics, finance, and game theory. The researchers find and analyze complex phenomena in their field and research tools analyzing complex phenomena are now developing.

In this article, we focus on complexity considerations in game theory. Game theorists accept the existence of complex problem in their field. Game theorists call the human behaviors in complex world "bounded rationality." For a long time, economists have treated human-beings as rational agents. What agents are rational is that they have all relevant information and computational power to solve any problem that they face. This is nonsense. Agents in real life don't have such information or computational power. Experimental studies of human behaviors show that agents in real life are often inconsistent with the behaviors that are assumed in ordinary economics.

Recently, the researchers in game theory focus on some complexity measures of automaton that implements agent's strategy or action. If the numbers of internal states of automaton is bounded on the limit that is determined by some complexity measure, then outcomes of the game are different from the case that all the agents are rational. We treat this problem of complexity measure in section 2.

In section 3, we focus on the problem of optimal mechanism design. Designing optimal economic institutions, for example, designing contracts, auctions, and tax payment, etc., is very important part of economic policy making or planning. Mechanism design is the general theory of these economic plannings. It is the problem in designing a particular mechanism that agents do not reveal their true preferences or information. Planner must take into account of incompleteness or indirectness in haviing players' preferences or information when she wants to design a mechanism. This is a hard task. Even if a mechanism designed, whether all the players in the mechanism can choose optimal choices is another problem. This problem leads us back to the complexity considerations in Game Theory described in section 2.

But we can find that bounded rational agents in the game can achieve optimal outcomes in learning process or evolutionary dynamics. For examples, optimal behaviors in auction mechanism can be achieved in the simulations with genetic algorithm and in experimental studies with human-subjects. These findings are described in section 4.

2. Games and Bounded Rationality

In this section, we show that there exists an optimal strategy but players cannot compute it. This example is called Diophantine Game by Jones[9]. This exampls show us that computational restriction on human behaviors or bounded rationality is needed when we should have reasonable outcome among agents in real life. Next, we see some complexity measures found in game theory.

2.1 Computability in Game Theory

In general, game treated in game theory are denoted by four tuples, $G = <N, S, C, P>$, where N denote the set of players, S the set of available strategies for players, C the set of outcomes, and P : $S^N \rightarrow C$ the set of payoff functions for the players.

Given a game, players choose their strategies to maximize their payoffs assuming that other players act rationally. These players' optimal choices of strategies are represented in some solution concepts. Nash equilibrium is the famous one.

Definition 2.1 [Nash Equilibrium]
Let S_i denote the strategy set for player i and S_j for other player j. A pair of $s_i^* \in S_i$, $s_j^* \in S_j$ is Nash Equilibrium if

$$\forall i, j \in N, \forall s \in A_i, \exists s^* \in A_i, \exists t^* \in A_j, \mu_i(s^*, t^*) > \mu_j(s, t^*)$$

Nash equilibrium is widely accepted by most economists and game gheorists as the most natural or useful solution concepts to analyze and describe economic consequences. In fact, at least one Nash equilibrium exists in any game within mixed strategy. So that we can use Nash equilibrium solution concepts for any situations.

In spite of usefulness of Nash equilibrium, we can show a game that no computable strategy exists even if a Nash equilibrium exists. In fact, Gilboa[4] and Ben-Porath[3] showed that under some conditions, calculating best response for other players' strategies in Nash equilibrium belonged to *NP complete problem*. NP complete problem is accepted as intractable problem in computer science (see Hopcroft & Ullman[8]). So we can say that if at least one equilibrium do exist, there exists the case that this equilibrium cannot be achieved among players with reasonable computational power, for example, polynomial time Turing Machine.

For example, Jones[9] showed a game named Diophantine Game in which there existed a winning strategy, but this strategy could not be computed.

Definition 2.2
Two players game $G = <S_1, S_2, \mu_1, \mu_2>$ is strictly competitive if

$$\forall s \in S_1, \forall t \in S_2, \ \mu_1(s, t) = - \mu_2(s, t)$$

where S_1, S_2 are the strategy set for player1 and player2, and μ_1, μ_2 are the payoff functions for player1 and 2 that $\mu: S \times S \rightarrow R$

Definition 2.3

Two players win-lose-games is a strictly competitive game G in which the range of payoff function is restricted in $\{0, 1\}$, where 1 means win and 0 lose.

$$\forall s \in S_1, \forall t \in S_2, \ \mu_1(s, t) \in \{0, 1\}$$

and

$$\forall s \in S_1, \forall t \in S_2, \mu_1(s, t) = 1 - \mu_2(s, t)$$

Definition 2.4

In two players win-lose-game, player i has a winning-strategy if there exists a strategy s that can satisfy $\mu_i(s, t) = 0$ for any strategy t of other player j.

Definition 2.5

A predicate $R(a_1, a_2, ..., a_m)$ is Exponential Diophantine if there exists a polynomial $P(a_1, a_2, ..., a_m, x_1, x_2, ..., x_n)$ with integer coefficients those operators are add, multiply and exponential and it satisfies that

$$R(a_1, a_2, ..., a_m) \Leftrightarrow \exists x_1, x_2, ..., x_n \ [P(a_1, a_2, ..., a_m, x_1, x_2, ..., x_n) = 0]$$

Definition 2.6 [Diophantine Game]

Diophantine Game is a win-lose game that thier payoff functions are Exponential Diophantine.

Theorem 2.1

In any two players win-lose-game, there exists a winning-strategy for eihter players.

Jones[9] showed a sequential move win-lose game with non-computable winning strategy in Diopahntine Game below. Player1 and 2 alternately choose an integer at each move .

First mover: player1
Second mover: player2
Player1's strategy set: $\{x_1, x_3, x_5\}$
Player2's strategy set: $\{x_2, x_4\}$
Payoff function for player1: $\mu_1(x_1, x_2, x_3, x_4, x_5) = x_1^2 + x_2^2 + 2x_1 x_2 - x_3 x_5 - 2x_3 - 2x_5 - 3$

In this game, player2 has a winning-strategy. Because if player1 has winning strategy, he can satisfy $\mu_1(x_1, x_2, x_3, x_4, x_5) = 0$, and then one can transform the polynomial, obtaining the equation

$$(x_1 + x_2)^2 + 1 = (x_3 + 2)(x_5 + 2)$$

But, player2 can choose her strategy x_2 that satisfies $(x_1 + x_2)^2 + 1$ to be prime for any x_1, player1 cannot satify $\mu_1(x_1, x_2, x_3, x_4, x_5) = 0$. Therefore he cannot win. In spite of existence of winning strategy for player2, player2 cannot choose her winning strategy for any time. Because whether there exists $n^2 + 1$ form prime number infinitely is an open problem in Number Theory(Guy[7]). Therefore player2 has a winning strategy but she cannot choose her winning strategy for any time.

2.2 Complexity Measure in Game Theory

Economists and game theorists often assume that all the players are rational, that is, they can solve any optimization problem. In the last section, we see the existence of the game with non-computable strategy in which there exists an equilibrium but it cannot be achieved among players with reasonable computational power. That is, players are not rational but their rationality are bounded on informational uncertainly or computability. This notion is called "bounded rationality" by H.A.Simon[13].

In this section, we argue some complexity measures in game theory. These complexity measures represent and model players' bounded rationality. We use *the prisoner's dilemma game* as an example in showing the differences of such measures. In the prisoner's dilemma game, players have two strategies, *cooperate*(C) and *defect*(D). Outcomes from their choices are represented in Figure.2.1. In this game, unique Nash equilibrium exists, and that is defect each other. But more profitable outcome exists, that is, cooperate each other. But this outcome is never appeared in Nash equilibrium among rational players. This is a dilemma.

	C	D
C	3, 3	0, 5
D	5, 0	1, 1

Figure 2.1 Payoff matrix of the Prisoner's Dilemma game.

If the prisoner's dilemma game is played repeatedly, two cases arise according to finite or infinite repetition. In finitely repeated game, unique Nash equilibrium is also defect each other in any time, that is shown using backward induction. In infinitely repeated game, cooperate each other is one of Nash equilibriums, but many other equilibria exist. That is, all the individually rational payoff that is higher than (1, 1) for both players are Nash equilibrium. This multiple equilibria problem is called "Folk Theorem"(see Figure 2.2).

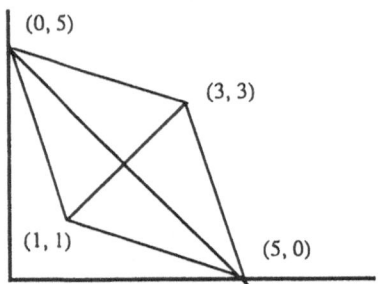

Figure 2.2 Payoff vector of Prisoner's Dilemma

In following sections, we argue how the models of bounded rationality with complexity measure resolve multiple equilibria problem described above.

2.2.1 Complexity as automaton implementing cost

Rubinstein[11] assumed that player selected his automaton implementing his strategy according to the following condition:

"First, maximize your payoff and minimize the numbers of states of automaton implementing your strategy."

Let $\pi_i(M_i, M_j)$ denote the payoff for player i when player i choose automaton M_i for implementing his strategy and player j M_j. And $|M_i|$ denote the numbers of internal states that M_i has.

Definition 2.7
Let $\pi_i(S, T)$ denote the limit of means of payoff stream for each player in the repeated games, where S and T denote automata which each player chooses.

Definition 2.8
Player i prefers an automaton M_i to any other automata M'_i avalalble for player i against player j's automaton M_j if

$$\pi_i(M_i, M_j) > \pi_i(M'_i, M_j)$$

or

$$\pi_i(M_i, M_j) = \pi_i(M'_i, M_j) \text{ and } |M_i| < |M'_i|$$

Definition 2.9
For any player i and automaton M_i for player i, a pair of automaton M^*_i and M^*_j for player i and other player j is Nash equilibrium if

$$\pi_i(M^*_i, M^*_j) > \pi_i(M_i, M^*_j)$$

or

$$\pi_i(M^*_i, M^*_j) = \pi_i(M_i, M^*_j) \text{ and } |M^*_i| < |M_i|$$

For example, under above lexicographic order criterion in the game with automata, any player prefers constant cooperate strategy to trigger strategy which cooperates unless the opponent defects and defects otherwise. Trigger strategy also achieve Nash equilibrium in repeated prisoner's dilemma game. Because players cannot deviate from cooperation on the equilibrium path, constant cooperate strategy can earn as same payoff as trigger strategy but which uses fewer internal states than trigger strategy has (see Figure 2.3).

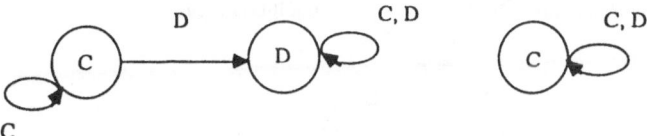

Figure 2.3 strategy implementing automaton. Left side is TRRIGER strategy that constructs a Nash equilibrium with each other, right side is ALWAYS COOPERATE strategy.

Under this lexicographic preference order, Rubinstein[11], Abreau & Rubinstein[1] showed that it appears more restricted outcomes in repeated prisoner's dilemma game than in the Folk Theorem. On Nash equilibrium with automaton, the equilibrium payoffs become convex combinations of {C,

C} and {D, D} or {C, D} and {D, C} (this payoff vector is shown as the cross in Figure 2.2).

2.2.3 Complexity as state switching cost

Another complexity measure for game strategy was suggested by Banks & Sundram[2]. This measure is that takes into account of cost of monitoring other players' actions or the frequency of switching the states of automaton. Players select his automaton minimize the frequency of switching the states of automaton according to the following condition:

"First, maximize your payoff and minimize the frequency of switching states of automaton implementing your strategy."

Under this criterion, defect each other is only Nash equilibrium in prisoner's dilemma game again. So we conclude that according to what complexity measure is applied, equilibrium outcomes differ. Problem of constructing more reasonable and useful complexity measure remains to future researches.

3. Mechanism Design

Economics of institutions is the research area of analyzing and designing optimal economic institutions. Designing contracts, auctions, and tax payment, etc., are popular cases. The general theory of designing economic institution is called *mechanism design* or *implementation problem*. In mechanism design, researchers have argued difficulty of designing optimal mechanism. And we will add complexity problem to these arguments.

Let N denote the set of players, P the set of preference profiles of players, S the set of strategy for players, and C the set of outcomes. Let F is the set of *Social Choice Function* $f : P \to A$, G the set of *Game Form* or *Mechanism* $g : S \to A$, and E the set of strategy select function $e : P \to S$ that selects strategies according to the set of players' preference profile P.

The problem of social planner is whether she can make a mechanism $g \in G$ whose set of outcomes is coincide with the outcomes selected in Social Choice Function F. The problem is described in Figuree 3.1.

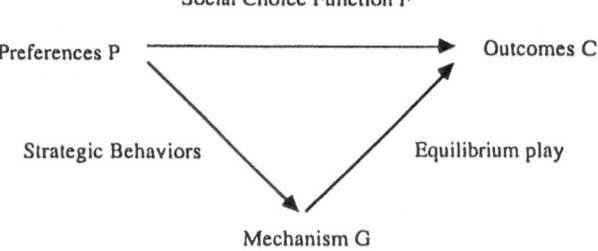

Figure 3.1 Concept of Mechanism Design

3.1 Difficulty in Mechanism Design

When a mechanism can be designed under Nash equilibrium solution concept, the mechanism is called *Nash Implementable*. Maskin[10] showed that if a choice function F was Nash-implementable, then it was *monotonic*.

Definition 3.1 [monotonicity]

A choice function F: P → A is monotonic if whenever c ∈ F(p) and c ∉ F(p') for any p, p' ∈ P, then there exists some player i ∈ N and some outcome b ∈ A such that c p_i b and b p_i c.

For example, we consider the problem in a biblical story, Wisdom of King Solomon(see Glazer & Ma[5]). Let two mothers claim the right to own a baby, Solomon's problem is designing optimal allocation mechanism that brings the baby to true mother. There are three possible outcomes:

  1. the baby is given to mother1
  2. the baby is given to mother2
  3. the baby is cut in two with sword

Preference relation for both mothers are described below.

$$p_1^{1}: a > b > d, \qquad\qquad p_1^{2}: a > d > b$$
$$p_2^{1}: b > d > a, \qquad\qquad p_2^{2}: b > a > d$$

Where p_1^{1} is mother 1's preference relation when mother 1 is true mother and p_1^{2} when mother 2 is true mother. p_2^{1} and p_2^{2} is likewise respectively.

Though King Solomon wanted to make a choice function, $F(p_1^{1}, p_2^{1}) = a$ and $F(p_1^{2}, p_2^{2}) = b$, it was *not* Nash-implementable, since it was not monotonic. For many situations, monotonicity is restrictive condition in machanism design.

Solomon's problem is fully resolved in *subgame-perfect equilibrium implementation*. Subgame-perfect equilibrium is the equilibrium that all the decisions among players are Nash equilibrium at any decision tree in extensive form game. This is a refinement of Nash equilibrium solution concepts.

Let mother1 must pay a fine $e > 0$ and mother2 M when they do not agree, where v_H is true mother's value for the baby and v_L is false mother's value, $v_H > v_L > 0$, and $v_H > M > v_L$. This mechanism is described in Figure 3.2. First, mother1 chooses *mine* or *hers*, then mother2 chooses *mine* or *hers*. Consider two cases:

case1. mother1 is true mother:

  If mother2 chooses *hers*, payoff for mother2 is $0 - 0 = 0$, else if mother2 chooses
  mine, payoff for mother2 is $v_L - (v_H + v_L) / 2 = (- v_H + v_L) / 2 < 0$ at subgame for
  mother2. So mother2 must choose *hers*. Then mother1 must choose *mine* at his
  subgame, because payoff for mother1 is $0 - 0 = 0$ when mother1's choice is *hers*
  and mother2's choice is *hers* and $v_H > 0$ when mother1's choice is *mine* and mother2's
  choice is *hers*.

case2. mother2 is true mother:

  If mother2 chooses *hers*, payoff for mother2 is $0 - 0 = 0$, else if mother2 chooses
  mine, payoff for mother2 is $v_H - (v_H + v_L) / 2 = (v_H - v_L) / 2 > 0$ at subgame for
  mother2. So mother2 must choose *mine*. Then mother1 must choose *hers* at his
  subgame, because payoff for mother1 is $0 - 0 = 0$ when mother1's choice is *hers*
  and mother2's choice is *mine* and $0 - e = - e < 0$ when mother1's choice is *mine* and
  mother2's choice is *mine*. .

Therefore, (*mine*, *hers*) = (1, 0, 0) and (*hers*, *mine*) = (2, 0, 0) are subgame-perfect equilibrium in this game form and resolve King Solomon's problem.

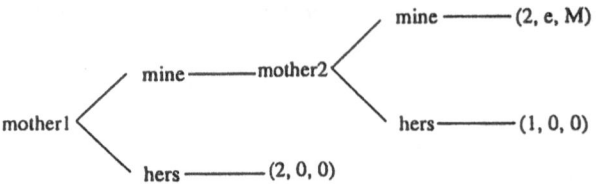

Figure 3.2 Solomon' s Problem under subgame-perfect equilibrium. Triplets at terminal node represent who
gets the baby, fine for mother1, and fine for mother2 in left-to-right order.

3.2 Complexity in Mechanism Design

Many economic decision problems are computationally complex. We already considered the problem of designing optimal mechanism for a specific problem. In this analysis, we assume that social planner has powerful computational power to solve any problem, but does not have preferences information of players' preferences fully. This means that if social planner has all the relevant information of players' preferences, he can design an optimal mechanism in any time. But this is impossible when we assume that social planner is also bounded rational agents.

Even though social planner can design an optimal mechanism, whether all the players can choose thier optimal choices in the mechanism is also another problem. We already looked that even though an equilibrium in a game existed, players couldn't choose optimal strategies in section 2. So there are two levels of complexity problems in mechanism design. First, whether we can design an optimal mechanism and second, whether all the players choose their optimal choices in the mechanism.

But reasonable model of mechanism design with bounded rational planner and players haven't constructed yet. It remains to future researches.

4. Towards Experimental Economics

When we want to have the optimal behaviors among players, we need not assume that all the players in the game are rational. Gode & Sunder[6] showed in their market experiments that regardless of participating a few pairs of players with random bidding strategies in the market, they could achieve efficient market price and quantity as well as in the market equilibrium with rational agents. This result is the most extreme case of relevance of bounded rationality. We need only to assume bounded rationality for solving the economic optimization problems. There is no need for rationality assumptions.

Next, we show in the computer simulations and in the experiments with human-subjects that players can learn optimal behaviors in the optimal auction mechanism, second price auction. These experiments show that human behaviors in real life are different with rational agents' behaviors, and those behaviors can achieve efficient outcomes as good as rational agents can do.

4.1 Auction Schemes

Consider optimal auction design problem and optimal behaviors among agents in the mechanism. There are some auction schemes widely used. Two auction scheme described below is compared in this section.

First Price Sealed Bid Auction
  Buyers put their bid in their envelopes and submit them, the highest bidder wins and pays his bid.

Second Price Sealed Bid Auction
  Buyers put their bid in their envelopes and submit them, the highest bidder wins and pays the highest bid among losers.

Let v_i denote buyer i's true value for the object to buy, $B_i(v_i)$ the bidding function and $r_i = max$ $s_j(j \neq i)$, where s_j is bid other than buyer i. We assume that buyer's true value is drawn from the uniform distribution. Then optimal bid in First Price Sealed Bid Auction is $(1 - 1/n)v_i$, though $s_i = v_i$ in Second Price Sealed Bid Auction. That is, it appears underestimate for the object to buy in First Price Sealed Bid Auction, though it appears true value revelation in Second Price Sealed Bid Auction. Second Price Sealed Bid Auction is the optimal mechanism in designing auction mechanism(see Vickrey[14]).

But above two auction schemes are equlivalent for the seller of the object.

Theorem 4.1 [Revenue Equivalent Theorem]
  Above two auction schemes achieve same expected revenue to the seller.

4.2 Can GA learn optimal strategy?

We used real value GA(genetic algorithm) to check whether bounded rational players could learn the optimal strategy for second price auction in evolutionary learning process.

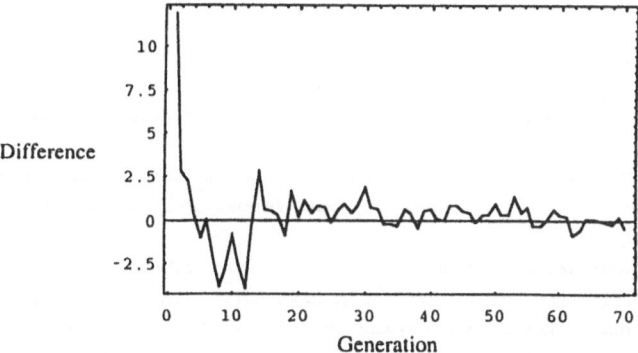

Figure 4.1 Simulation result of learning optimal strategy in second
price auction using real value GA.

Suppose any strategy has a linear strategy ax + b for his true value x for the object to buy. Parameters x, a and b are drawn from the uniform distribution those rages are $x \in [50, 150]$, $a \in [0, 2]$ and $b \in [0, 25]$. Population has 50 individuals, we gather 5 individuals per a group and they compete in each group. And we iterate this process 200 times, then select better performance agents by means of roulette selection rules. After selecting agents for next generation, two GA operators, crossover and mutation, are applied to the agents where crossover_rate ▬ 0.5 and mutation_rate ▬ 0.01.

We examined whether real value GA could learn theoretical value (1.0, 0.0). The results we got were represented in Figure 4.2. We concluded that GA could learn optimal strategy in Second Price Sealed Bid Auction.

4.3 Experiments of Second Price Auction

We also conducted the experiments of Second Price Sealed Bid Auction with human-subjects. In this experiments, we got some significant characters of players' expectations or actions described below that didn't appear in rational agents or GA simulations.

(1) Some players deviated from equilibrium play at the end of repetition of the Game.

(2) Many players cared about their ranking in auction.

(3) Players played Nash equilibrium when they felt that they won, but they wished to reduce other players' payoff when they felt that they lost.

(1) is known as the end behaviors. (2) is not appeard in Nash equilibrium. (3) is observed in other experiments, for example, that of providing public goods voluntarily. Saijo[12] called this phenomenon "spite". We think that the spite behaviors of players are related with pivotness of the mechanism. Pivotness is the character that benefit is determined by player's own decision, but its costs is determined by other players' choices. Second Price Sealed Bid Auction is the case. If the mechanism has pivotness, players maximize their payoff when they feel that they win, but minimize other players' payoff when they feel that they lose. This is spite behaviors.

Surprisingly, when there exists spite behaviors, outcomes can also become efficient. For example, selfish spite behaviors among players can achieve cooperative outcomes in providing public goods(Saijo[12]). Thus, human behaviors in real life is different with rational agents behaviors, but they can be efficient. Constructing more realistic behavioral model of human behaviors remains future researches. This task can be achieved in collabolations with theoretical stusies and experimental researches.

References

[1] D.Abreu & A.Rubinstein(1989) "The structure of Nash equilibrium in repeated games with finite automata", *Econometrica* 56, p.1259-1281

[2] J.Banks & R.Sundaram(1990) "Repeated games, finite automata, and complexity", *Games and Economic Behavior* 97, pp.97-117

[3] E.Ben-Porath(1990) "The complexity of computing a best-response automata in repeated games with mixed strategies", *Games and Economic Behaviour* 2, p.1-12

[4] D.K.Gode & S.Sunder(1993) "Allocative efficiency of markets with zero-intelligence traders", *J.Politicl Economy* 101, pp.119-137

[5] I. Gilboa(1988) "The complexity of computing best-response automata in repeated games", *J. Economic Theory* 45, p.342-352

[6] J.Glazer & C.A.Ma(1989) "Efficient allocation of a 'prize' - King Solomon's dilemma", *Games and Economic Behavior* 1, pp.222-233

[7] R. K. Guy(1981) *Unsolved problem in number theory*, Springer-Verlag

[8] J.E. Hopcroft & J.D.Ullman(1979) *Introduction to automata theory, language and computation*, Addison-Wesley

[9] J. P. Jones(1982) *"Some undecidable determined games"*, *International J. Game Theory* 11, p.63-70

[10] E.Maskin(1977) "Nash equilibrium and welfare optimality", MIT mimeo.

[11] A. Rubinstein(1986) *"Finite automata play the repeated prisoner's dilemma"*, *J. Economic Theory* 39, p.83-96

[12] T. Saijo (1995) "The spite dilemma in voluntary contribution mechanism experiments", *J.Conflict Resolution* 39, pp.535-560

[13] H. A. Simon(1955) "A behavioral model of rational choice", *Quaterly Journal of Economics* 69, p.99-118

[14] W.Vicrey(1961) "Counterspeculation, auctions, and competitive sealed tenders", *J.Finance* 16, pp.8-37

Forecasting and Stabilizing Economic Fluctuations Using Radial Basis Function Networks

Taisei Kaizoji, (International Christian University),
Toru Hattori, (Central Research Institute of Electric Power Industry)

Abstract

The purpose of this study is to estimate structural equations for an economic system from observed data and to find a stabilization policy for the economy. We employ a simple Keynesian (IS-LM) model to represent the economy, which generates (chaotic) time-series data of GNP and interest rate. We will approximate the structural equations of this economy using Radial Basis Interpolation method [4] and stabilize the economy based on the approximation.

Key words : a Keynesian model, endogenous business cycles, chaos and RBI method.

1. Introduction

Time series of many macroeconomic variables have the appearance of being irregular fluctuations. The traditional explanation of this is that an essentially stationary economy is subject to random shocks.

An alternative way of treating irregular fluctuations would be to use a model involving non-linear difference (or differential) equations. If a model of this sort would exhibit chaotic or cyclical behavior, then it could provide an explanation for the noisy fluctuations[1].

In this paper we employ the following dynamic Keynesian (IS-LM) model, which has often been studied as one of the basic model of endogenous business cycles.

$$Y(t+1) = F(Y(t), r(t)) = Y(t) + \alpha[C(Y(t)) + I(Y(t), r(t)) - Y(t)] \tag{1}$$

$$r(t+1) = G(Y(t), r(t)) = r(t) + \beta[L(Y(t), r(t)) - \bar{M}] \tag{2}$$

where $Y(t)$ is GNP in time t, $r(t)$ is interest rate in time t, and $\bar{M}$ is fixed money supply. F and G represent excess demand for commodities and money respectively. Let us define other parameters and functions as follows :

Adjustment speed of good market : $\alpha = 0.3$
Adjustment speed of money market : $\beta = 0.6$
Consumption function : $C(Y(t)) = 5 + 0.7Y(t)$
Investment function : $I(Y(t), r(t)) = 10 + 0.1Y(t) + (12 - 0.4Y(t))r(t)$
Money demand function : $L(Y(t), r(t)) = 0.5 \exp(1.2(4 - r(t))Y(t)$

The economic equilibrium (Y^*, r^*) is $(33.9443, 5.2044)$ under these parameter setting, and the eigenvalues of Jacobian matrix evaluated at (Y^*, r^*) are $\lambda_1 = 0.3001$ and $\lambda_2 = -1.8646$.

---

[1] As the reader can see from [2], there exists a rich population of economic models that produce nonlinear dynamics, and empirical results which are suggestive of significant non-linearities and chaos for a variety of economic time series.

These verify that the economic equilibrium (Y^*, r^*) is unstable, and furthermore the economic fluctuations become chaotic[2].

2. Stabilization Policy

Since in general the unstable endogenous business cycle is considered to have negative effects on economic activity, the government is expected to stabilize the economy as a policy goal. Suppose that the government wishes to use money supply to achieve the equilibrium interest rate r^*, and also wishes to ensure that the monetary policy stabilizes the economy. We consider the following simple discretionary monetary policy rule.

$$M_{t+1} = \bar{M} + \delta(r_t - r^*). \tag{3}$$

The logic behind this policy is simply to increase money supply wherever interest rate is above its target level r^*, and to decrease it when it is below its target. Considering the feedback control rule of money supply (3) in conjunction with the IS-LM model (1) and (2), we obtain the three simultaneous difference equations. The necessary and sufficient condition for the local stability of the equilibrium (Y^*, r^*) is easily obtained as

$$1.42427 < \delta < 1.63368. \tag{4}$$

Note that the stability condition (4) is effective in the neighborhood of the equilibrium. Thus the government should implement monetary policy according to (3) when GNP and interest rate at time t is closely approaching the equilibrium level, (specifically, $|Y_t - Y^*| < \epsilon$ and $|r_t - r^*| < \epsilon$). If the government chooses the appropriate parameter which satisfies the condition (4), then the monetary policy can stabilize the chaotic business cycle and make the economy converge to the equilibrium[3].

3. Approximation of the Structural Equations by RBF Network

So far we have implicitly assumed that the government knows the economic structure. In fact, however, the government must somehow infer such a structure from observed data. The present study will approximate economic structure represented by equations (1) and (2) by using the RBI method, which recently attracting attentions as a method to approximate unknown nonlinear functions from observed data.

3.1. Radial Basis Interpolation Method [4]

Given (Y_t, r_t, M_t), $t = 1, 2, \ldots, T$, the relationship between (Y, r) in t and (Y, r) in $t + 1$ will be described by using nonlinear functions F and G as $Y_{t+1} = F(Y_t, r_t)$, and $r_{t+1} = G(Y_t, r_t, M_t)$. However, since functions F and G are unknown, we need to approximate them by other functions f and g. The RBI method is a method of approximating nonlinear functions using the weighted sum of radial basis functions ψ as follows :

$$y = \sum_{j=1}^{N} \lambda_{hj} \psi(\| x - x_j \|) \tag{5}$$

where λ_{hj} is the j th factor of the vector of weights $(h = Y, r, M)$, y is the output vector $((Y, r)$ at $t + 1$ here), and x is the input vector $((Y, r, M)$, in t here.) x_j is called the RBF center, N of which are located in the possible intervals. The radial basis function is a function such that output level depends only on the distance between the input vectors and RBF center,

---

[2]For a full account of the dynamic properties of the model, see [1].
[3]For more information about controlling chaos, see [1] and [3].

which is considered as a type of network and often called the Radial Basis function Network. We adapt the Gaussian function as a specific functional form of the radial basis function, $\psi(\| \boldsymbol{x} - \boldsymbol{x} \|) = \exp(- \| \boldsymbol{x} - \boldsymbol{x}_j \|^2 / b)$. The vector of weights λ_j, $(j = 1, 2, ..., N)$, which are only unknown in (5), is estimated by a least squares method. Using the estimated λ_j, we are able to approximate the structural equations F, G.

3.2 Stabilization based on the Approximation

We used observed data as the RBF center to approximate the functions. The equilibrium calculated from the approximated functions is $(Y^*, r^*) = (33.948, 5.2)$ which is almost equal to the equilibrium of the model (1) and (2). Furthermore the necessary and sufficient condition for the local stability of the equilibrium (Y^*, r^*) is

$$1.2086948 < \delta < 1.59111.$$

Comparing them with (4), it is obvious that the range of possible policy parameters considerably coincides. Therefore it is possible to estimate the appropriate parameters for monetary policy from the approximated structural equations. Figure 1 shows the stabilization effect when we set $\delta = 1.5$, $\epsilon = 0.1$.

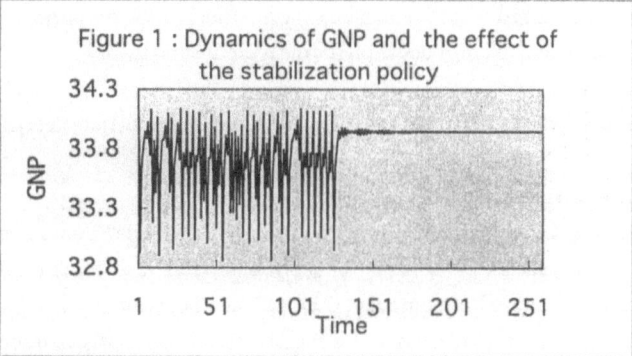

Figure 1 : Dynamics of GNP and the effect of the stabilization policy

4. Concluding Remarks

It is likely that the actual data involves various types of noise that affect business cycles. In this case, we need to identify the noise and separate from the effect explained by the structural equations. The conventional RBI method is not appropriate if the data contain the noise. Then, a corrected RBI method needs to be considered.

References

[1] T.Kaizoji, C-S. Chang, Chaotic Business Cycles and the Stabilization Policy in a Dynamic Keynesian Model, forthcoming to *The Journal of Social Science*, 1997, Vol.35.

[2] J. Benhabib, ed., *Cycles and Chaos in Economic Equilibrium*, Princeton University Press, Princeton/New Jersey, 1992.

[3] E. Ott, T. Sauer, J. A. Yorke, eds., *Coping with Chaos : Analysis of Chaotic Data and the Exploitation of Chaotic Systems*, A Wiley-Interscience Publication John Wiley and Sons, Inc. 1994.

[4] T. Poggio, F. Girosi, Networks for Approximation and Learning, *Proc. of the IEEE*, Vol.78, No.9,1990, pp.1481-1497.

Parallel Processing Economic System Composed of Individuals under Satisfying Principle

Masaaki Yoshida[1]

1 Dept. of Economics Senshu Univ., Tama-ku, Kawasaki-shi, Kanagawa, 214-80, JAPAN,
 E-mail: VAH29311@pcvan.or.jp

Abstract

 If we treat time as an irreversible flow, then neoclassical economics which interpret our economy as an equilibrium of behaviors of optimizing agents cannot serve even an approximation. It is because that the neoclassical equilibrium system cannot be attained in the case that each agent makes irreversible decision independently as in our actural economy. To comprehend such an economy appropriately, we constructed a parallel processing economic model composed of agents under satisfying principle in Keynes' text and made a simulation system. This system serves a foundation to analize our economy as a complex system, and in addition provides a technical foundation for IO-analysis and a micro-foundation for the reproduction theory.

Key words: Complexity, irreversible time, satisfying principle, neural network, Keynes

1. Introduction

 How should we represent a complex system in economics? For long time economists has looked individuals in their model as utility or profit optimizers, and construct their economic model as an equilibrium composed of consistent decisions and behavior of these individuals. But in the actual economy, time flows irreversibly. So if each decision under disequilibrium is made independently, the tatonnement process cannot work and the equilibrium cannot be attained. Moreover in our world, each decision must be made in a short time. But optimizing calculation process under constraints is NP-class problem, so such a behavior cannot be executed in a finite and meaningful time. When the object of making economic model is to comprehend how our economy behaves, such a way to study seems very strange. But economists supported such models because of lack of alternatives. Contrary to the tradition of economics, J.M.Keynes, especially in his *Treatise on Money*, substantially developed very unique theoretical system appropriate to express such an economy. This paper attend to reconstruct Keynes' text and show a consistent alternative system. We use the satisfying principle in turns of the optimizing principle to formulate behaviors of individuals, and use the neural network based parallel processing model in turns of the ordinary market equilibrium model.

2. Starting points in Keynes

 Over 50 years have passed after Keynes' death, it is still hard to find a general agreement about the theoretical structure of his economics. Many economists have tried to reconstruct the keynesian economic theory so many times from *General Theory* , but this appealing book doesn't show its whole structure easily. Some of it is because of its confusions, but the main reason is that Keynes himself did not make up his discussions with the ordinary neoclassical framework. If you trace the path of his development of thought, from *Tract on Money* to *General Theory*, it is not so difficult to find a consistent flow in it. Keynes started only with a simple monetary quantity equation and without the general equilibrium framework in *Tract on Money* , but because of the inability of monetary quantitiy theory to explain economic fluctuations, he tried to expand his monetary economic model with *working capital, step by step method, bearishness function, division of production sector into consumption and capital goods* ... etc. A compilation of such efforts is *Treatise on Money*. In the chapter of fundamental equation, two patternized processes are mentioned. One is of entrepreneurs which adjust their output-level upward or downward according to extra profit. The other is of households which adjust their consumption level upward or downward according to their income change. The output of households directly affects the amount of consumption, and the output of entrepreneurs affects the employment level, so it affects the income level. Therefore if we combine consumption goods-producing entrepreneurs process and household process together and change the output-level of capital goods producing entrepreneurs, an easy simulation will show us the multiplier process. In this book there are more several processes of agents, and Keynes' discussion is founded on simulations of interactions of these processes.

To understand Keynes' discussion appropriately, two steps are important in reconstruct his theoretical framework,

(1) Setting up patternized decision process model as observed in our economy,

(2) Design of input-output relation of each process.

(More about the historical part, please see my papers, 'Theoretical Structure of Treatise on Money', Nov. 1988, in *Keizai-Ronso*, Kyoto Univ., 'Keynes' Treatise on Money as an Integrated-Process System', Oct. 1991, in *Senshu Keizaigaku Ronshu*, Senshu Univ.)

3. Formulation

We saw a ground-design for a non-neoclassical economics above, then to see this system has a concrete theoretical style and enough handlablility, let's construct our model on the base of Hopfield-type neural network. Now we get our fundamental equation for keynesian system,

$$X_j(t+1)=a_j f[\Sigma_{d=0} \Sigma_{i=1} w_{ji}(t-d)X_i(t-d) + z_j(t) - s_j - \Sigma_{d=0}s'_j(t-d)X_j(t-d)]$$

where,

$X_j(t+1)$ adjustment output of process j at t period,

w_{ij} weight of output of process i as input of process j, if there is no effect then wij=0,

s_j satisfying level of process j

z_j stimulation to process j from outside of the system

a_j parameter of maximum adjustment value for process j.

f[ ]   threshold function, which products positive or negative constant values according to the sign of [ ]

In our economy, we can observe several groups whose behavior show similar pattern, for example, entrepreneur of consumption goods, entrepreneur of capital goods, wealthy class, household of laborer, etc. Similar behavior pattern individuals combined into a random connected network which can be treated by statistical dynamics. As macro-behaviors of each groups are also described same type threshold functions, now we get a parrarel processing economic system composed of macro groups who show patternized behaviors.(About theoretical detail, see, 'Who killed 'Keynes'?,' Mar. 1993, in *Annual Report of The Inst. of Socialscience Res.of Senshu Univ.*)

3. Simulation

To illustrate the behavior of our economic system, we made a simulation system. We have no room to show the whole algorithm here, but next figure shows how things are going in simple case. If you only see processes marked * , it is easy to see a micro-foundation of the principle of effective demand as a special case. As you see, the system has the same structure of Hopfield network and has many equilibrium according to its parameters. As economic implications, it shows

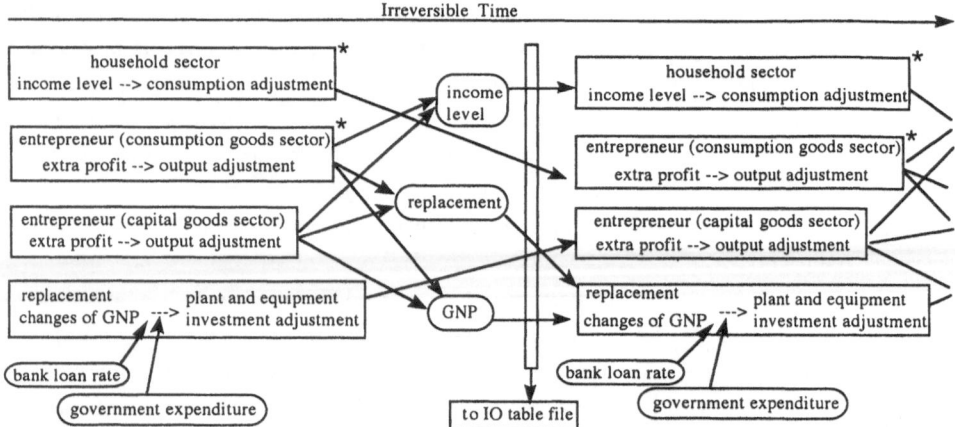

how our economic system behaves in irreversible time and how it holds some robustness. Additionary , our program shows the technical relation of each processes and IO table, and provides the micro-foundation of a reproduction schema as an IO table at one of equilibriums.

4. Conclusion

Our system is constructed with the intention to comprehend our actual economy. So we started with satisfying agents, and went on constructing a social model on the base of a mutually connected type neural network model. Our system needs the data about what input level is normal. It may seem ad hoc, but we cannot imagine a newcommer without any idea about the normal way in his bisiness world. The world which our system intent to express is such a actual world.

Learning is not mentioned in our report, but it is not so difficult to expand the system to include learning with introducing a multi-layer control model into each decision process. With such a version-up, our model can describe an institutional change as a change of equilibrium map. But it should be noted that such changes take much more time than each action of individual. Contrary to the world of game theory, our system includes many irreversible time dimensions.

In this system, any behavior of agent is described by threshold functions, so the whole system, which is described by a composite function of many non-linear functions, is so complicated that we cannot treat its solution path analitically, that is, a complex system.

Model of public city image formation based on a positive feedback mechanism

Shigeru Tanaka[1] and Yoshiko Tanaka[2]

1 The Inst. of Physical and Chemical Res. (RIKEN), FRP, Wako, Saitama 351-01, Japan
2 Life Design Institute, Chiyoda-ku, Tokyo 100, Japan

Abstract
In the present work, we first construct a model of the formation of public images of a city based on the idea that public city images are selected from variety of individuals' city images according to the distribution of individuals' image preferences and that positive feedback from the public images to the individuals' images accelerates the sharing of a city image among people. The model is formulated in terms of population dynamics. An equation derived equation to model the populations of two city images competing with each other is applied to the analysis of the results of our questionnaire survey on the images of a city that is well-known in Japan as a prison city. Comparing the theoretical results and data obtained from the survey, we suggest a new planning method for city revitalization.

Key words: City image, positive feedback, city planning

1. Introduction
In Japan, hardware-oriented planning has greatly contributed to city revitalization by improving the social infrastructure. Recently, however, difficulties in this type of planning have often been encountered due to the movement of production bases to developing Asian countries, the lack of funding in local governments for the maintenance of facilities and the recent cessation of increasing tax revenue [1]. To overcome these difficulties, we previously proposed image-mediated planning as a substitute for conventional hardware-oriented planning [2]. The proposed type of planning is based on the idea that positive feedback from 'public images' [3] to individuals' images through intra- and intercity communication accelerates the sharing of a specific city image among the residents, which promotes the establishment of a city identity strongly associated with the residents' 'we-feeling'. Image-mediated planning facilitates city revitalization at the grass-roots level beginning with residents' attitudes, as opposed to hardware-oriented planning, which is likely to be conducted by a public sector in a top-down fashion. In this study, we examine the positive feedback effect on the process by which a particular city image becomes shared by the residents, analyzing statistical data obtained from our questionnaire survey on images of Abashiri, a famous prison city in Japan, on the basis of a mathematical model which we propose here.

2. Model description
When people living in a city are continually exposed to the city's real features, such as natural resources, landscape, historical background and culture, they come to hold images of the city in their minds, which are insider images. Likewise, people living outside the city can hold some images regarding the city, which are outsider images. These images interact with one another through personal and mass media communication, with the result that the population with respect to particular images changes. It is surmised that residents of a city tend to have common images because of a general tendency that people want to hold opinions in common with a dominant portion of the members of their reference group. Therefore, most images initially evoked by representing the city's real features in individuals' minds may be eliminated by selection, and some images become shared by all the residents. This tendency is accelerated as the sharing of insider images progresses, because an image that is commonly held by more residents attracts people more strongly. In this way, some of the initially evoked individuals' images grow into public images that are commonly possessed by the residents and can be observed as typical images of the city. If the process of sharing images is regarded as feedforward in city image formation, the attractive force from public images to individuals' images can be looked upon as positive feedback. The formation of insider public images is schematized in Fig. 1 [4].

Note that the functional role of positive feedback in the process of sharing city images is quite similar to that in the formation of public opinions [5]. Therefore, we attempt to construct a mathematical model for the segregation of public city images from various individuals' images, according to Weidlich and Haag's formulation.

Suppose that two images are competing in population with each other in a group composed of N persons, which is divided into subgroups composed of n^k persons depending on the difference in image preference between the two images, αk. In addition, n_1^k and n_2^k are assumed to represent the subpopulation in which members support images 1 and 2, respectively. The public image strength is defined by $x = \sum_k \left(n_1^k - n_2^k\right)/N$, which takes 1 when all members of the group support image 1, and -1 when they all support image 2. Because of space limitations, we omit the detailed formulation of our model, and just show the ensemble-averaged equation describing the population dynamics of images,

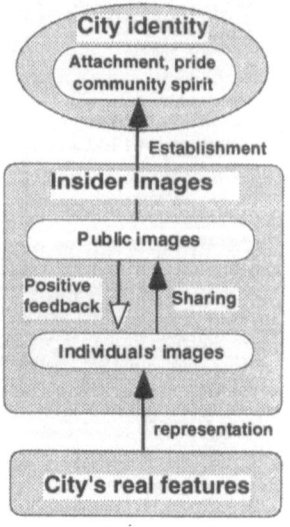

Figure 1

$$\frac{dx^k}{dt} = -x^k + \tanh(\alpha k + \beta x) . \ (1)$$

βx on the right-hand side represents a positive feedback effect of the public images on the growth of images within the subgroups specified by k. When we define f^k as $f^k = n^k/N$, the public image strength x obeys the equation

$$\frac{dx}{dt} = -x + \sum_k f^k \tanh(\alpha k + \beta x) . \ (2)$$

Obviously, the fixed point solution x_0 to Eq. (2) is given by $x_0 = \sum_k f^k \tanh(\alpha k + \beta x_0)$. If this model well describes the public city image formation, $\tanh(\alpha k + \beta x_0)$ as a function of k should be given by $g^k = \left(n_1^k - n_2^k\right)/\left(n_1^k + n_2^k\right)$ which is computable from the statistical data from a questionnaire survey. The values of α and β can be determined by this fitting procedure.

3. Analysis of the results of a questionnaire survey
To better understand the basic mechanism for the public city image formation by applying the proposed model, it would be convenient to work with a city which has two dominant city images, so we selected Abashiri as the site for a questionnaire survey, as it is well-known in Japan for two main images: a prison located in the city, and icebergs which float to its seashore on the Sea of Okhotsk every winter. We analyzed these images, in particular. Questionnaires were distributed to 800 residents of the city, who were selected at random from the resident register, and 1,000 outsiders through 125 chambers of commerce and industry. The major questions were as follows: (1) To what extent do you think 'icebergs' and 'prison' adequately describe Abashiri? Check one of five levels of adequacy for each. (2) Which do you support as an image of Abashiri, 'icebergs' or 'prison'? The former question is meant to evaluate people's inherent preference between the two images; the latter is meant to examine populations in which members support 'icebergs' and 'prison'. We defined images 1 and 2 as "icebergs" and 'prison', respectively, when analyzing the data. The details of the survey methods are described in reference 4.

Figure 2 shows f^k and g^k for images held by the insiders and outsiders. By fitting g^k with $\tanh(\alpha k + \beta x)$, we obtained the values of α and β for the insider and outsider images: $\alpha = 2.85$ and $\beta = 1.04$ for insiders ($N = 220$); $\alpha = 4.90$ and $\beta = 0.93$ for outsiders ($N = 805$). $g(x) = \sum_k f^k g^k(x)$ calculated using these values are depicted for insiders and outsiders in Fig. 3. From the intersection points of $y = g(x)$ with $y = x$, we obtain $x_0 = 0.26$ for insider images and $x_0 = 0.71$ for outsider images, which are close to the values of public image strengths obtained from the survey 0.29 and 0.77, respectively. The slope of $y = g(x)$ around $x = x_0$ for residents of Abashiri is almost 1, and the intersection with the x-axis is close to the origin. These facts indicate that the state of the public images is in the vicinity of the critical point between the monostable and bistable states. Although the public image is biased toward 'prison' at the present, it would require only a slight shift in residents' preferences to change 'icebergs'. For outsider images, since the slope of $y = g(x)$ is less than 1, the state of the public images is monostable. The intersection point of $y = g(x)$ with the x axis is shifted greatly to the right, which implies that the outsiders' preference is strongly biased toward 'prison'.

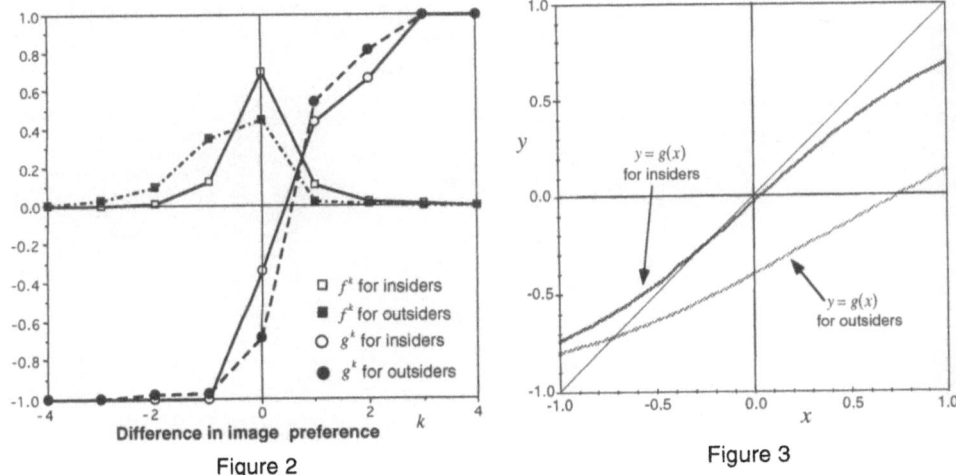

Figure 2

Figure 3

Generally, image preference is hard to change to a large extent, because it is a sort of 'attitude' which is held firmly by people for a long time. Particularly, the outsider image of 'prison' seen in the above analysis has been consolidated and seems unchangeable. Accordingly, it is unlikely that planning to increase the number of visitors to Abashiri by changing the outsider images from 'prison' to 'icebergs' will be successful. On the other hand, it is expected that a representative image among public insider images will become the core of the city's identity when the image is linked to residents' positive feelings about their city, such as attachment to the city and pleasure in living there. City identity thus established produces a 'we-feeling' which promotes, in turn, coherent and cooperative actions by the residents, and, as a result, the city is revitalized. In our data analysis, we found a strong correlation between a residents' preference for the 'iceberg' image and positive feelings regarding Abashiri (data not shown). Taken together, we would propose a planning method for Abashiri to change its public images from 'prison' to 'icebergs' by encouraging the residents' preference for 'icebergs'. This image-mediated planning will be successful in that 'icebergs' will tend to become the core of the city identity, and hence the city identity will promote city revitalization developed from the residents' minds.

4. Conclusion
First, the present research based on a combination of a mathematical model and a questionnaire survey demonstrates that images can be measured quantitatively, despite the fact that they are conventionally regarded as ambiguous and intractable. Second, excellent agreement between the data from the survey and the mathematical model indicates that public images emerge through the positive feedback effect that people want to hold common images with others. Finally, it is suggested that image-mediated planning making the best of insider images would be more effective than utilizing outsider images in the case of Abashiri.

References
[1] Y. Tanaka and Y. Kumata. Synergetic effects of planning and self-organization in city identity creation, The Korean Journal of Regional Science 1 1 99-109, 1995.
[2] Y. Tanaka and S. Tanaka. Positive feedback model for city vitalization, Int. J. Japanese Sociology, 5 107-122, 1996.
[3] K. E. Boulding. *The Image*. The University of Michigan Press, 1956.
[4] Y. Tanaka. Studies of the formation of city images, Ph.D. thesis, Tokyo Institute of Technology, 1996.
[5] W. Weidlich and G. Haag. *Concepts and Models of a Quantitative Sociology: The Dynamics of Interacting Populations*. Springer-Verlag, Heidelberg, 1983.

Index of Contributors

Key Word Index